谨以此书献给在中国召开的第22届国际博物馆协会大会

Dedicated to the 22nd General of the International Council of Museums in China

中国博物馆建筑

MUSEUM BUILDINGS IN CHINA

中华人民共和国国家文物局指导
BIAD传媒《建筑创作》杂志社编

天津大学出版社
TIANJIN UNIVERSITY PRESS

序

无论以前人们曾经有过多么辉煌的文明基业，都无一例外地被历史的尘埃所湮没。但是，我们仍然可以通过某个途径去寻找这些文明的踪迹，这个途径就是博物馆。长期以来，国际博协与中国在各专业领域的交流与合作富有成果，特别欣喜的是三年一届的国际博协大会将于2010年11月首度落户中国上海。此次大会将把国际博协和中国乃至亚洲的博物馆领域之间的合作推向一个崭新的水平。

2010年5月18日，“国际博物馆日”在中国上海世博中心举办，国际博物馆协会会旗首次在上海世博中心东门升起。本届世博会期间国际博协以“博物馆，城市之心”为主题，在6个月的展期内分别举办欧洲、非洲、北美洲、拉丁美洲及加勒比地区、大洋洲和亚洲专题月活动，全方位展示世界多样化的文化自然遗产和丰富的博物馆资源，从文化、经济、创新、社会、环境五个方面诠释博物馆和城市之间互相依存的关系。2010年上海世博会的主题是“城市，让生活更美好”，与第22届国际博协大会的主题“博物馆与和谐社会”相得益彰。无论是主题演绎，还是高科技展示技术、文化领域的国际合作等都具有共同关注的命题。

此次由国际博协2010年大会筹委会及中国国家文物局组织配合推出的《中国博物馆建筑》一书是从中国3 000余家博物馆中精选项目的集萃，除故宫博物院等中国古代建筑精华之作外，均选取近年来中国博物馆建筑创作上的佳作，尽展中国博物馆建筑的最新成果，把有特色的中国博物馆介绍给世界。我们从中可以看到，博物馆建筑空间不仅是一种“人性尺度”的空间，更是一种人类与历史沟通的媒介。面对日益更新的、具有永恒时代精神的场所，博物馆建筑更将成为承载人类智慧的“精品”。

在此，我祝愿《中国博物馆建筑》一书可以成为博物馆界永久经典的专业参考书目。

国际博物馆协会主席　亚力桑多·康明斯

2010年11月

Preface

All civilisations, no matter how glorious, will one day be buried in history. Luckily, however, these civilisations can still be discovered in institutions such as museums. For a long time, ICOM (International Council of Museums) has been making real progress in its exchanges and cooperation with heritage professionals in China. In addition, the organization's General Conference, which takes place every three years, will be held in Shanghai for the first time ever, in November 2010. This event will enhance ICOM's cooperation with the Chinese and even Asian museum sectors.

On 18 May, 2010, International Museum Day was celebrated in the Shanghai Expo Center, where the ICOM flag was raised for the first time at the Eastern Door. During the six-month-long Expo, ICOM held six monthly exhibitions to highlight the regions of Europe, Africa, North America, Latin America and the Caribbean, Oceania and Asia on the theme of *Museums, Heart of the City.* ICOM's objective at the Expo was to present the world's diversified culture and natural heritage, to demonstrate the substantial resources held by museums, and to make a lasting statement about the interdependent relationship between museums and cities under five headings: culture, economy, innovation, society and environment. The theme of the Shanghai Expo 2010, *Better City, Better Life,* and that of the 22nd General Conference of ICOM *Museums for Social Harmony* are complementary. The World Expo and ICOM have many common concerns, including theme interpretation, high-tech exhibition skills and international cooperation in the cultural field, etc.

Museum Buildings in China, a publication compiled by the organising committee of the 2010 ICOM General Conference and the State Administration of Cultural Heritage, presents a fine selection of museums chosen from more than 3,000 museums in China. Besides pillars of Chinese ancient architecture, such as the Palace Museum, this book also highlights more recent masterpieces, demonstrating the latest achievements in Chinese museum architecture. It also reinforces the view that architectural space is not only a space for "humanity", but also a bridge connecting human beings with their history. The eternal and ever-evolving nature of museum highlights their role as some of the most important carriers of human wisdom.

I hope that *Museum Buildings in China* will be a reference tool for museum professionals long into the future.

ICOM President Alissandra Cummins

November, 2010

序

公共建筑体现着一个城市的文化底蕴和精神，在建筑设计师眼中，作为公共建筑的博物馆，不仅是普通的楼宇，更是传承文明、延续历史、沟通文化的精神圣地。建筑设计师们在设计博物馆时，往往不仅要全力展示本时期、本地域的建筑科技水平，而且要努力与历史、文化相结合，并充分发挥艺术想象力，尽可能地使博物馆建筑经得住历史的考验，成为一座经久不衰的“未来历史建筑”。

当前，中国博物馆事业正处于一个前所未有的高速发展时期。在党和政府的高度重视下，中国的博物馆数量持续快速增长，一大批功能齐全、具有重要影响和象征意义的新馆相继建成，并成为一些地方的标志性建筑。不仅有由两院院士吴良镛担纲设计、与博物馆历史建筑组合和谐的南通博物苑新展馆，也有“弃宏大而就无形”的陕西汉阳陵博物馆；不仅有代表中国本土化设计精品的安阳殷墟博物苑，也有中外合作设计的博物馆建筑杰作，包括中国电影博物馆、中国科技馆新馆等。我们高兴地看到，越来越多的建筑师日益重视中国的博物馆建筑设计，关注和参与博物馆的建筑设计项目，越来越多的博物馆建筑正以其引人注目的设计而形成了超越地域的文化影响力。

为了系统总结近百年来中国博物馆建筑的发展历程，全面展示中国博物馆建筑的创作成就，在国际博物馆协会第 22 届大会召开前夕，在中国国家文物局、国际博物馆协会第 22 届大会筹备委员会和各地文物部门、博物馆的支持下，北京市建筑设计研究院《建筑创作》杂志社和天津大学出版社共同编撰推出了《中国博物馆建筑》一书。该书从中国现有的 3 000 余座博物馆中精选了在建筑方面具有代表性的博物馆 100 家，对其建筑特色进行专题介绍，力求通过建筑师、城市规划师之视点传达 21 世纪博物馆建筑的中国发展趋向。

在《中国博物馆建筑》一书杀青付梓之际，我谨代表国家文物局和国际博物馆协会第 22 届大会筹备委员会，向参与本书编撰工作的北京市建筑设计研究院《建筑创作》杂志社和天津大学出版社编辑同人们表示热烈的祝贺，向为本书出版提供大力支持的故宫博物院等 78 家博物馆表示诚挚的谢忱。我相信，这本图文并茂的出版物将带给世界博物馆界一个惊喜，因为通过它大家可以品味各具特色的中国博物馆建筑，感受东方文化及中华文明的深邃魅力。

祝愿国际博协第 22 届大会在中国成功举办!

中华人民共和国国家文物局 局长　单霁翔

2010 年 11 月

Preface

A public building embodies the spirit of the city's cultural heritage, and in the eyes of architects, as a public building' the museum is not only the ordinary building, but also the spiritual holy land for inheriting civilization, continuing history and commucating culture. When designing a museum, designers often should not only fully display the technological level of construction at that time, in that area, but also work hard to integrate history and culture, give full play to artistic imagination, and make the museum withstand in the test of history as much as possible and become an enduring "future historical building".

At present, Chinese museums are in an unprecedented period of rapid development. The party and the government attach great importance to sustained and rapid growth of Chinese museums, a large number of new functional and symbolic museums have been built, and became landmarks in some places. There are not only Nantong Museum's new exhibition hall which is designed by the academician of the Chinese Academy of Sciences and China Academg of Engineering Wu Liangyong and with harmonious combination of historical buildings, but also Shaanxi Hanyangling Museum which concentrated on "invisible rather than magnificent"; and not only Anyang Yin Ruins Museum which represented the Chinese local design style, but also masterpieces with Chinese-foreign cooperation, including China National Film Museum, the new China Science and Technology Museum' and so on. We are pleased to see that more and more architects increasingly pay attention to the design of Chinese museums, concern and get involved into the architectural design projects of museums, and also more and more of the museum buildings with striking designs are forming cultural influence beyond the geography.

In order to systematically sum up the museums in China over the past century and give a comprehensive display of the creative achievements of the Chinese museums, a book named "Museum Building in China" was edited by "Architectural Creation" magazine in Beijing Institute of Architectural Design, together with Tianjin University Press on the eve of the 22nd General Conference of the International Council of Museum, which is also in support of the State Admini stration of Cultural Heritage, the Preparatory Committee of the 2010 ICOM General Conference and local heritage sectors. The book selected 100 museums with representative architectural designs among 3 000 Chinese museums, and made special introduction trying to convey the Chinese development trend of the 21st Century Museum through the point of view from architects and urban planners.

Before the Publication of this book, I, on behalf of the Chinese State Admini stration of Cultural Heritage and the Preparatory Committee of the 2010 ICOM General Conference, want to express our sincere gratitude to the compilers from the "Architectural Creation" magazine in Beijing Institute of Architectural Design and Tianjin University Press, and say thanks to the more than 100 museums that made huge support for the book. I believe that this illustrated publication will bring the museum world a surprise, because through it we can taste different characteristics of the Chinese museums, feel the oriental culture and the deep charm of Chinese civilization.

Wish ICOM 22th General Conference will be successfully held in China!

Director General of the State Administration of Cultural Heritage of the People's Republic of China Shan Jixiang

November, 2010

《中国博物馆建筑》编委会

执行编辑部

本书策划： 金　磊

执行编辑： 李　沉　王　珅

本书撰文： 李拱辰　王　珅　黄希明　刘卫华　许俊德　史金龙　杜　松　李亦农　柯　蕾　陈韶峰　张炽康　谢晓婷　陆　军　李照斌　郑栋梁　任　巍　苏　雷等（除署名外，文稿均由各博物馆提供）

美术编辑： 胡珊瑚　黄　岽　张墨贤

英文翻译： 许松华　水润宇　苗　淼　王　祥　魏霖霖（除署名外，文稿均由各博物馆提供）

英文校审： 李广欣

编　　务： 冯桂红

建筑摄影： 刘锦标　杨超英　陈　鹤　邢荣哲　闫　冬　李绪成　李　升　王　飞　张　惠　张益帅　厉晋春　张　鹏　李拱辰　牛爱红　唐伟超　贲道春　王　琼　李乐年　李国伟　舒　赫　莫修权　麦　西　北京泛道国际建筑设计咨询有限公司（除署名外，照片均由各博物馆提供）

特别鸣谢

基于近年来中国博物馆建设迅猛发展，本书在编撰中恐有遗漏，真诚望业内外同道予以谅解。在此，我们要感谢各地文物局在遴选项目上给予的帮助，并特别鸣谢如下单位的鼎力支持：

故宫博物院　中国国家博物馆　中国人民革命军事博物馆　中国科学技术馆　中国地质博物馆新馆　中国农业博物馆　中国航空博物馆　中国铁道博物馆　中国现代文学馆　中国电影博物馆　首都博物馆　北京自然博物馆　北京天文馆　周口店北京人遗址博物馆　房山世界地质公园博物馆　天津博物馆　河北省博物馆　西柏坡纪念馆　山西博物院　八路军太行纪念馆　内蒙古博物院　“九一八”历史博物馆　旅顺博物馆　吉林省博物院　吉林省自然博物馆　黑龙江省博物馆　铁人王进喜纪念馆　上海博物馆新馆　中国航海博物馆　上海科技馆　南京博物院　侵华日军南京大屠杀遇难同胞纪念馆新馆　苏州博物馆　南通博物苑　扬州博物馆（中国雕版印刷博物馆）　常州博物馆　无锡博物院　浙江省博物馆　中国丝绸博物馆　温州博物馆　安徽省博物馆　福建博物院　中国闽台缘博物馆　江西省博物馆　八一南昌起义纪念馆　山东新博物馆　青岛市博物馆　中国甲午战争博物馆　河南博物院　郑州博物馆　洛阳博物馆　中国文字博物馆　湖北省博物馆　荆州博物馆　毛泽东遗物馆　广东省博物馆　深圳博物馆　广西民族博物馆　柳州博物馆　海南省博物馆　重庆中国三峡博物馆　广汉三星堆博物馆　邓小平故居陈列馆　金沙遗址博物馆　建川博物馆　贵州省博物馆新馆　西藏博物馆　陕西历史博物馆　秦始皇兵马俑博物馆　延安革命纪念馆　汉阳陵博物馆　西安博物院　甘肃省博物馆　青海省博物馆　新疆维吾尔自治区博物馆　香港历史博物馆　澳门特别行政区海事博物馆　澳门艺术博物馆

目录

Contens

The Palace Museum

故宫博物院

故宫博物院是在明清两代沿用近 500 年的皇家宫殿范围内建立的以宫殿建筑为载体、宫廷史迹与皇家收藏并重的综合性国家博物馆。它位于北京城的中心地段、皇城历史文化保护区内，总占地面积约 1 060 000 m²（含紫禁城、午门至端门地段、大高玄殿、皇史宬、内务府御史衙门 5 处地段）。

故宫又称紫禁城，占地逾 720 000 m²，是中国古代宫城发展史上现存的唯一实例和最高典范，也是世界上现存规模最大、保存最完整的古代宫殿建筑群。1961 年 3 月由国务院公布为第一批“全国重点文物保护单位”，1987 年 12 月由联合国教科文组织列入《世界文化遗产名录》。城垣四面辟门，建有城楼，四隅建有重檐多脊、四面显山的角楼，周围绕以 52 m 宽的护城河。紫禁城内文物建筑的房屋基址占地面积 157 577 m²，分为外朝和内廷两大部分。南为外朝，北为内廷。外朝以太和殿、中和殿、保和

The Palace Museum is housed in the former imperial palace of the Ming (1368-1644) and Qing (1616-1911) dynasties. With its historical sites and a vast imperial collection, the Palace Museum is regarded as one of the most comprehensive museums in China. Situated in the center of downtown Beijing and within the Imperial City Historical Conservation Area, the heritage site of the Palace Museum today covers a total of 1,060,000 square meters including the Forbidden City and four adjacent areas: area from Meridian Gate to Origination Gate, Hall of Grand Profundity, Imperial Archives and Censorial Agency of Imperial Household Department.

The Palace Museum, also known as the Forbidden City, is the only existing example of China historical imperial palace with the largest scale, highest rank and most completely preserved architectural complex in the world. In March 1961 the Palace Museum was designated by the State Council as one of China's foremost-protected cultural heritage sites, and in December 1987 was made a UNESCO World Heritage site. The Forbidden City covers an area of 720,000 square meters, with a 52-meter-wide moat surrounded on four sides. Four pierced gates face four directions and a watchtower with a double-eave roof covered with yellow glazed tiles stands at each corner. The footprint of the architectural complex inside the Forbidden City is 157,577 square meters, spreading out on either side of an invisible central axis. The southern portion is known as the Outer Court, centers on three main halls — Hall of Supreme

01

01. 铜缸局部

对页

02. 太和殿

03. 角楼剖面

01. Part of a bronze Vat

Opposite page

02. Hall of Supreme Harmony

03. Section of the turret

02

殿为中心，用于举办朝会大典；内廷以乾清宫、交泰殿、坤宁宫为主体，用于日常生活起居。除朝寝殿宇外，还有辅佐皇帝理政的衙署、祭祀所用的佛堂道场以及用于娱乐休闲的园林和戏台。

现有陈列大致分为宫廷原状陈列、宫廷史迹专题陈列、宫廷艺术品专馆陈列、临时展览几种。宫廷原状陈列和宫廷史迹专题陈列主要分布于外朝和内廷的主要殿宇及两庑。绘画、陶瓷、珍宝等宫廷艺术品专馆陈列分布于外朝两侧的文华殿、武英殿及内廷东侧的宁寿宫，临时展览则布置在午门、神武门城楼的专用展厅内，构成完备的展示体系。

Harmony, Hall of Central Harmony, and Hall of Preserving Harmony. It was here in the Outer Court that the emperors held court and conducted grand ceremony. While the northern portion is the Inner Court, including the Palace of Heavenly Purity, the Hall of Union, and the Palace of Earthly Tranquility. The Inner Court is comprised of not only the residences of the emperors and Their consorts but also venues for religious rituals and administrative activities.

The present exhibitions in the Palace Museum mainly devote to historic imperial life, imperial history, imperial art and temporary exhibitions. The imperial life and imperial history exhibitions are mainly displayed in the halls and corridor rooms of the Outer Court and the Inner Court. The Hall of Martial Valor is paintings and calligraphy gallery, the Hall of Literary Brilliance is ceramics gallery and the Palace of Tranquil Longevity is treasure gallery, the Gate of Divine Prowess and the Meridian Gate are specially arranged for temporary exhibitions.

03

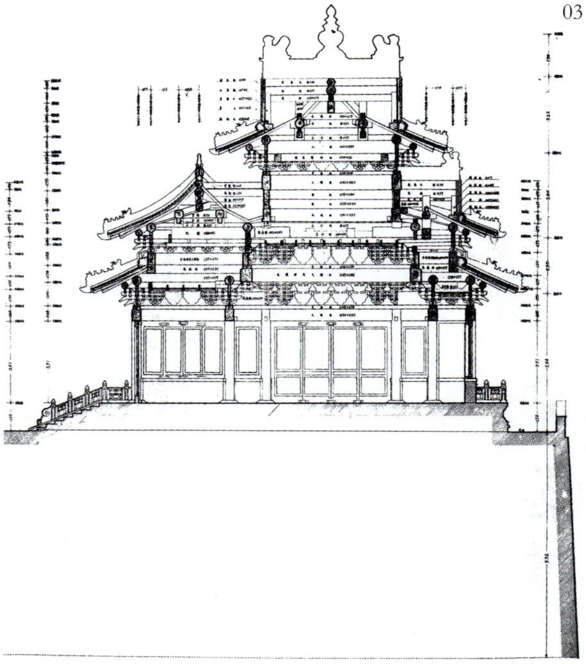

04-05. 故宫远眺　　04-05 Distant view of the Palace Museum

06

07

08

09

08. 故宫角楼 08.Turret of the Palace Museum
09. 角楼晨景 09.Turret in the morning

对页 **Opposite page**
06. 中和殿 06.Hall of Central Harmony
07. 保和殿 07.Hall of Preserving Harmony

National Museum of China

中国国家博物馆

中国国家博物馆位于天安门广场东侧，与人民大会堂遥相呼应。该馆于 2003 年 2 月 28 日在原中国历史博物馆和中国革命博物馆两馆合并的基础上组建成立，隶属于中华人民共和国文化部，是以展示历史与艺术并重，集收藏、展览、研究、考古、公共教育、文化交流于一体的综合性国家博物馆。

中国历史博物馆的前身为 1912 年 7 月 9 日成立的“国立历史博物馆筹备处”。1949 年 10 月 1 日，在中华人民共和国成立的同日，更名为“国立北京历史博物馆”；1959 年更名为“中国历史博物馆”。中国革命博物馆的前身为 1950 年 3 月成立的“国立革命博物馆筹备处”；1960 年正式命名为“中国革命博物馆”。 1959 年 8 月，位于北京天安门广场东侧的两馆大楼竣工，为建国十周年十大建筑之一。同年 10 月 1 日，在国庆十周年之际，两馆开始对外开放。

中国国家博物馆坚持“以人为本”的建设发展理念，以“贴近实际、贴近生活、贴近群众”为原则，坚持“与我们这样一个

The National Museum of China (NMC) was established on February 28, 2003 and is directly under the Ministry of Culture of the People's Republic of China. Formed on the basis of the former National Museum of Chinese History and National Museum of Chinese Revolution, the NMC primarily focuses on history and art and aims at displaying the splendours of China's history , culture and so on.

The National Museum of Chinese History could be traced to the Preliminary Office of National History Museum founded on July 9,1912. The predecessor of the National Museum of Chinese Revolution was the Preliminary Office of National Museum of Revolution established in 1950. The main buildings for the two museums, located on the east side of Tian'anmen Square in Beijing, was completed and opened to the public in 1959.

"The National Museum" of China insists to pursue the development concept of "Putting Human First", with maintaining "close contact with reality, life and the public" as the principle and persists in the direction of foundation such as "to be commensurate with the standing of a country such as ours, Chinese nation with long history and brilliant civilization, vigorous development of socialist modernization and the increasing needs of spiritual culture of the people." Aiming at the top international standards and leading the development of the museums in China.

The National Museum of China is a museum with the largest gross floor area in the world. It will be playing an important and positive role in protecting national cultural heritage, showing our country with long history, promoting Chinese civilization, developing foreign-oriented cultural exchanges and embodying soft power of

01

01. 老馆南入口

对页
02. 总平面
03. 新馆外景

01. South entrance of old museum

Opposite page
02. General layout plan
03. Exterior of new museum

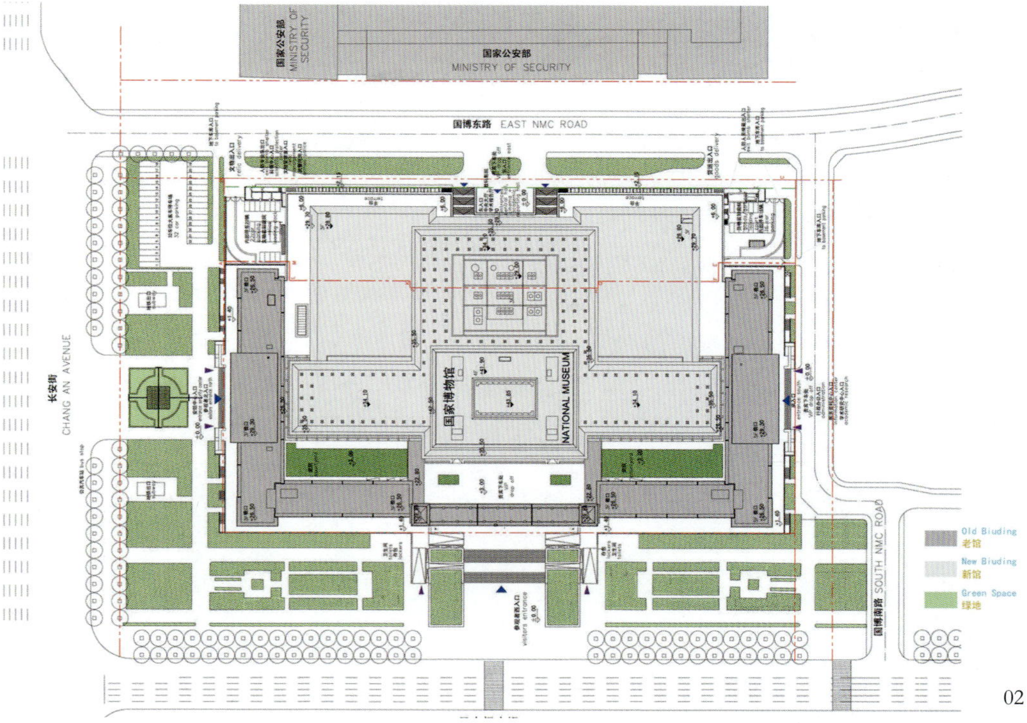
国家公安部
MINISTRY OF SECURITY
国博东路 EAST NMC ROAD
长安街
CHANG AN AVENUE
国家博物馆
NATIONAL MUSEUM
国博南路 SOUTH NMC ROAD
参观者西入口
visitors entrance
Old Biuding
老馆
New Biuding
新馆
Green Space
绿地

02

03

04. 天安门广场整体鸟瞰

04. Aerial view of Tian'anmen Square

对页
05. 新馆西立面效果图

Opposite page
05. Building volume of the NMC and the buildings around

大国地位相称，与中华民族悠久的历史和灿烂的文明相称，与蓬勃发展的社会主义现代化事业相称，与广大人民日益增长的精神文化需求相称”的建馆方向，以“国内领先、国际一流”为建馆目标，坚持“人才立馆、藏品立馆、业务立馆、学术立馆”的办馆方针。

中国国家博物馆是世界上建筑面积最大的博物馆，将会在保护国家文化遗产，展示祖国悠久历史、弘扬中华文明、进行爱国主义教育、开展对外文化交流、体现中华文化软实力等方面发挥积极而重要的作用。为适应构建公共文化服务体系和建设学习型社会的需要，该馆将成为广大公众特别是青少年学习历史和文化知识、接受文明熏陶、进行终身学习的文化阵地和课堂。

2007 年 3 月至今，中国国家博物馆进行了改扩建工程，馆舍总建筑面积 191900 m^2，硬件设施和功能达世界一流。藏品数量达 100 余万件，展厅数量 49 个，设有“古代中国”、“复兴之路”两个基本陈列，还设有十余个各艺术门类的专题陈列展览及国际交流展览。

Chinese culture, etc. The museum will be a cultural place for lifelong learning which the public, especially teenagers will study history, culture knowledge and receive civilization here in order to meet the requirements of constructing the public support system and building the learning society.

From March, 2007, the NMC was under a renovation and enlargement project. Today, the new NMC has a gross floor area of 191,900 square meters with forty-nine galleries and more than one million collections. The public could visit two permanent exhibitions: “Ancient China” and “The Road of Rejuvenation”. Also there are more than ten special art exhibitions and international exchange exhibitions on Show.

05

06

07

06. 老馆南门厅
07. 南老馆办公层

6. South hall of the old museum
7. office floors of the old south museum

08. 新馆入口大厅效果
09. 新馆标准展厅效果

8. Rendering of the new museum's entrance hall
9. Rendering of the new museum's standard exhibition hall

Military Museum of the Chinese People's Revolution

中国人民革命军事博物馆

中国人民革命军事博物馆是"国庆十大工程"之一，位于海淀区复兴路北侧，占地 80 000 m²，建筑面积 60 557 m²。主体建筑平面呈"山"字形，地下 1 层，地上中央大厅 7 层，上部塔座 2 层，塔顶上为军徽，总高 97.47 m；东西两侧地上 4 层，两翼地上 3 层，设有大小陈列室 10 个。中央大厅的后部是三层的连接厅和一层带走马廊的大跨度拱顶的兵器馆。

中央大厅贯穿两翼陈列室、序幕厅、交通厅、兵器馆及上层的中心，是博物馆主要入口。大厅为 30 m×30 m 的方形厅，共二层，中间为 18 m×18 m 的回廊空间，绿色实心大理石栏板，中间 12 根浅绿色大理石柱子，配以嵌有金星的汉白玉柱头。大厅两侧为交通厅，设有楼梯及电梯。

中央大厅两侧及两翼陈列室基本布局为桶形，开间 8 m，进深三跨 24 m。展览路线以每一个翼一层为一个段落，采用"大口袋"的布置办法，顺序循环路线。

Military Museum of the Chinese People's Revolution is one of the "Ten Great Buildings for National Day", located on the north of Fuxing Road, Haidian District, covering 80,000 square meters, with a floor area of 60,557 square meters. The plane of the main building looks like the Chinese character "山". There is one floor underground, seven floors overground for the central hall, with two pedestals of the upper layer. On the top of the tower There is an army emblem. The altitude of the building is 97.47 meters. There are four floors overground on the east and west sides, three stories on the two wings, ten exhibition rooms in different sizes. At the back of the central hall lies a three-layer hall for connection and a one-floor weapons exhibition hall with a big arch span.

The central hall through the exhibition halls, the prologue halls, the transport halls and the weapons exhibition hall on the wings, and the upper center, is the main entrance of the museum. The lobby is a square hall of 30m×30m, and the middle is the cloister of 18m×18m by space, with green solid marble tailgate. In the middle there are twelve light green marble columns, which are equipped with white marble chapiters paneled with gold stars. On both sides of the hall are transport halls, setting stairs and lifts.

The basic arrangements of the central hall and the exhibition halls on the wings are the shape of the barrel, 8 meters wide, 24 meters deep with 3 spans. Each floor of a wing is an independent part of the tour route. The circulation route

01

01. 博物馆旧影

对页
02. 鸟瞰
03. 中部正立面

01. Old photograph of the museum

Opposite page
02. Aerial View
03. Central facade

02

博物馆的外立面采用渐次升高的造型，外墙勒脚为花岗岩蘑菇石，假石墙面。所有入口均为花岗石，嵌有石刻旗帜花纹和带有勋章图案的柱头门罩，屋顶檐头饰黄色琉璃砖，建筑造型雄伟稳重。竖立在中央塔顶上的军徽是军事博物馆的主题标志。博物馆全部为钢筋混凝土框架体系，中央顶部军徽的塔筒为钢骨架、钢筋混凝土筒壁。

is clockwise and the way of arrangement is like a big pocket.

The facade of the museum has a figure gradually increasing. Outer plinths are made of granite mushroom stones and the walls are made of false stones. All the entrances are made of granite, which have carved flags inlayed and stigmas with a decoration of medals. The roof is dressed with yellow glazed bricks. The whole building is magnificent and steady. The army emblem at the central top of the tower is the theme of the museum. The museum is reinforced concrete framework system. The tower at the top is steel reinforced, with reinforced concrete around.

03

04

05

06

04-06. 室内
04-06. Interior

China Science and Technology Museum

中国科学技术馆

中国科学技术馆是我国唯一的国家级综合性科技馆。其主要教育形式为展览教育，通过科学性、知识性、趣味性相结合的展览内容和参与互动的形式，反映科学原理及技术应用，鼓励公众动手探索实践，不仅普及科学知识，而且注重培养观众的科学思想、科学方法和科学精神。

中国科学技术馆新馆位于北京奥林匹克公园中心区，西濒奥运水系，南依奥运主体育场，北望森林公园，占地 48 000 m^2，建筑规模 102 000 m^2，是奥林匹克公园中心区体现“绿色奥运、科技奥运、人文奥运”三大理念的重要组成部分。

以“体验科学，启迪创新，服务大众，促进和谐”为建馆理念的中国科学技术馆新馆，在工程建设和内容建设的全过程中始终贯穿科学发展观的主线，注重“世界眼光，中国特色”。新馆设有“科学乐园”、“华夏之光”、“探索与发现”、“科技与生活”、“挑战与未来”五大主题展厅，公共空间展示区及球幕影院、巨幕影院、动感影院、4D影院等4个特效影院。此外，

With exhibition education as its main form of education and through hands-on exhibition contents that combine science with knowledge and interest, China Science and Technology Museum presents scientific principles and their applications, encourages the public to learn science through exploration and practice in person. This way, the museum not only popularizes scientific knowledge, but also accentuates the cultivation of visitors' scientific thinking, scientific methods and scientific spirit.

Located within the common domain of the Beijing Olympic Green, China Science and Technology Museum borders the Olympic water system in the west, is adjacent to the Beijing National Stadium (better known as the Bird's Nest in the south), and faces the Forest Park in the north; it covers an area of 48,000 square meters and has a floor area of 102,000 square meters. In essence, the museum constitutes an important component of the common domain of the Beijing Olympic Green, which embodies the concepts of "Green Olympics, High-tech Olympics and People's Olympics".

With "facilitating science experience, inspiring innovation, serving the general public and promoting harmony" as its founding philosophy, China Science and Technology Museum carried out the construction and content development of its new premises in line with the main line of the scientific outlook of development in the entire process, emphasizing "global vision and Chinese characteristics". The new museum features

01

01. 球幕影院

对页
02. 新馆夜色
03. 总平面
04. 首层平面

01.The dome screen theater

Opposite page
02. The new premises of China Science and Technology Museum at night
03. Master plan
04. Plan of ground floor

02

N
39000
156000
4层
H=41.16
5层
H=44.55
143000
东入口广场
西入口广场
南入口广场
总平面图

03

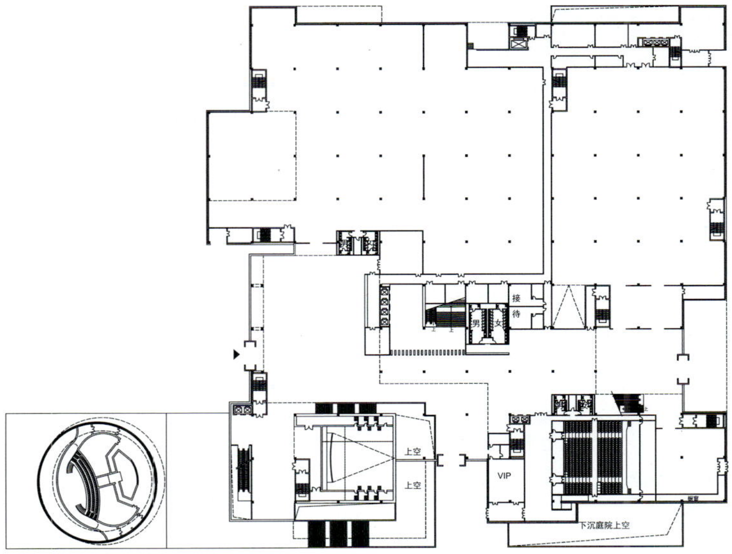
接
待
男
女
上空
上空
VIP
下沉庭院上空

04

05. 西立面

对页
06. 立面
07. 剖面

05.The west side of the architecture

Opposite page
06. Elevation
07. Section

05

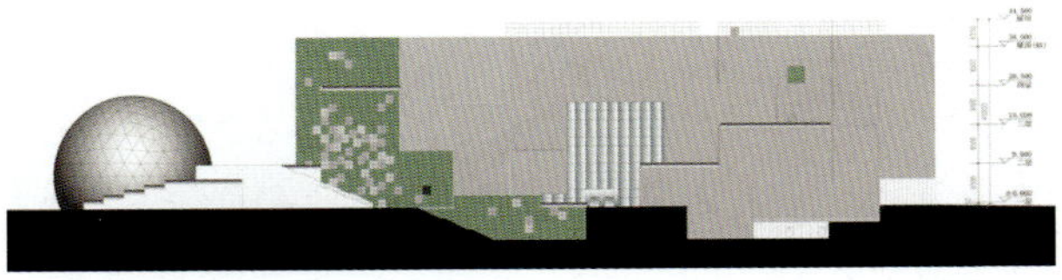

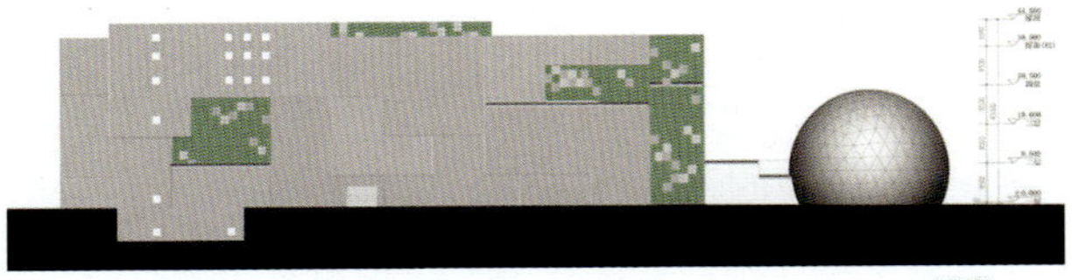

06

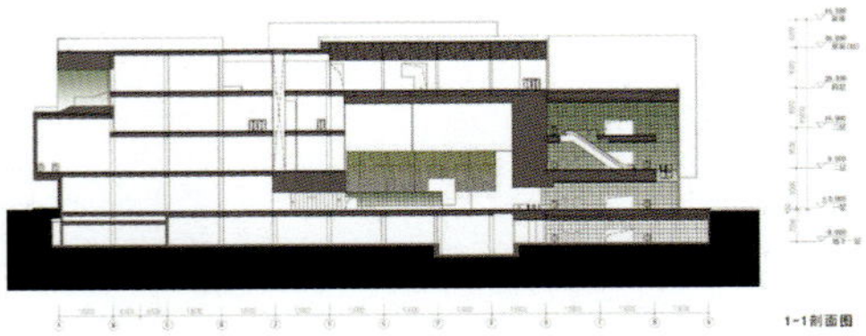

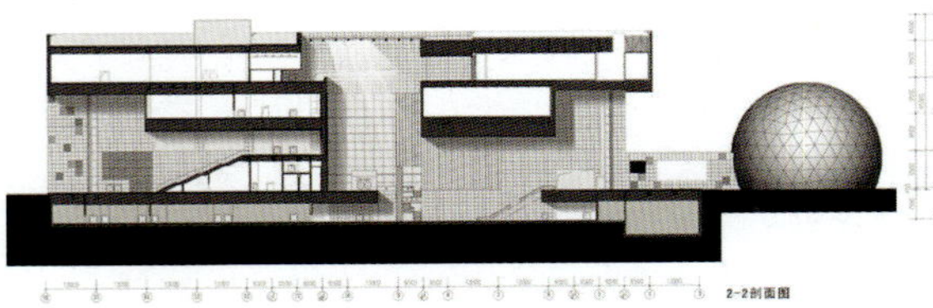

07

08

新馆设有多间实验室、科普报告厅及多功能厅。

新馆建筑为一体量较大的单体正方形，外形设计为若干个积木般的块体相互咬合，使整个建筑呈现为一个巨大的“鲁班锁”，又像一个“魔方”，蕴涵着“解锁”、“探秘”的寓意。

five theme exhibition halls, namely, “Science Paradise”, “The Glory of China”, “Exploration and Discovery”, “Sci-Tech and Life” and “Challenges and the Future”; it also has several public space display zones and four special-effect theaters, i.e. a dome screen theater, a giant screen theater, a motion theater and a 4D theater. In addition, the new museum boasts a number of popular science laboratories, a popular science lecture hall and a multifunctional hall.

A gigantic single construction project, the contour of this cubic-shaped architecture of the new museum is composed of a number of interlocking building blocks, making the whole structure appear like a huge “Lu Ban Lock” as well as a magic cube, symbolizing “unlocking” and “exploration”.

09

08. 球幕影院 08. The dome screen theater
09. 中央大厅 09. The central lobby

对页 **Opposite page**
10. 动感影院 10. The motion theater
11. 巨幕影院 11. The giant screen theater

10

11

12

13

12. “华夏之光” 展厅之中国古代天文仪器
13. “探索与发现” 展厅之“宇宙之奇” 展区

12. The ancient astronomical instruments of the Glory of China theme exhibition gallery
13. “The Wonder of the Universe” zone of the Exploration and Discovery theme exhibition gallery

14

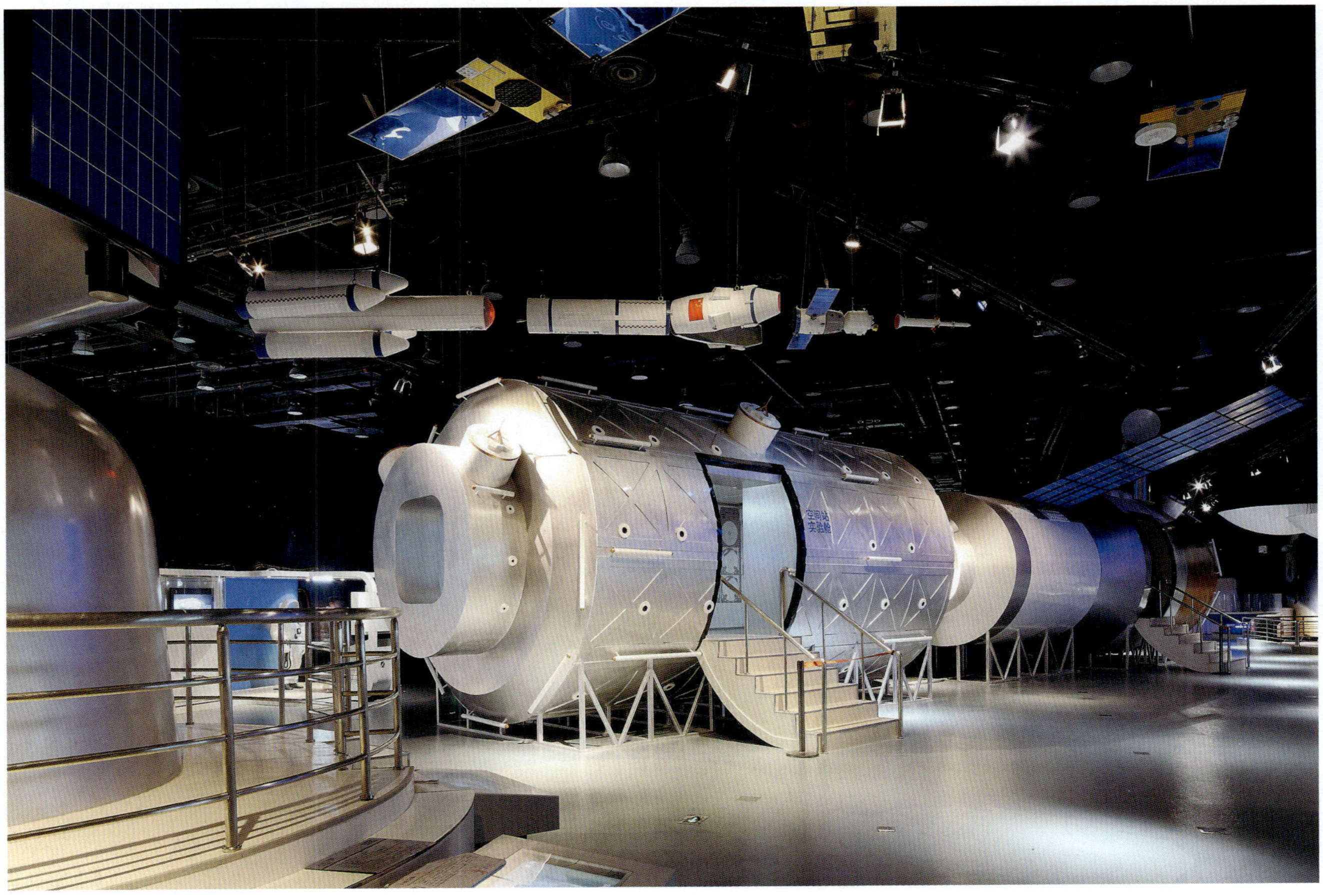

15

14.“科技与生活”展厅之“交通之便”展区
15.“挑战与未来”展厅之“太空探索”展区

14. “The Convenience of Transportation” zone of the Sci-Tech and Life theme exhibition gallery
15. The “Space Exploration” zone of the Challenges and the Future theme exhibition gallery

The Geological Museum of China

中国地质博物馆新馆

中国地质博物馆是国际上久负盛名的地质学博物馆，始建于1916年，前身为中央地质调查所地质陈列馆，是中国第一个国立自然科学博物馆，也是我国历史最悠久的博物馆之一。

中国地质博物馆大楼现位于北京西四街，建筑面积11 000 m²，1958年建成，1959年10月对社会开放，一直是西四地区的标志性建筑。

2004年完成的中国地质博物馆修缮改造工程是建馆以来规模最大、最为彻底的一次改造工程，包括大楼修缮加固、内部安全防范功能完善等硬件工程和管理运营等软件方面的提升。

中国地质博物馆经过修缮改造，展陈面积达到4 500 m²（含观众休息区）；同时全面创新展陈手段，大量采用声光电、多媒体、虚拟现实等技术，重建了地球厅、矿物岩石厅、宝石厅、史前生物厅、国土资源厅5个基本陈列展厅和1个临时展厅，生动地展示了国土资源国情、地球科学原理和精彩的标本世界。

The Geological Museum of China, which was built in 1916, is an internationally famous museum. It used to be the geological exhibitions of Central Geological Survey. It is the first national science museum of China, and also one of the oldest domestic museums.

The Geological Museum of China is located on Xisi Street, Beijing, with a floor area of 11,000 square meters, completed in 1958, and has been open to the society since October, 1959. It has been a landmark building in Xisi area.

The renovation and reformation of the Geological Museum of China which was completed in 2004 is the largest and the most complete ever since. It included reinforcement of building repairs and improvement of internal security. It also included promotion of management and operations.

The space for exhibition in the Geological Museum of China has reached 4,200 square meters since the renovation and reformation(rest area included). The means of exhibition and display are full of innovations. The technology of sound, light, electricity, multimedia and virtual reality has been used a lot. Five basic halls includ the Earth hall, the minerals rock hall, the gemstone hall, the prehistoric creatures hall and the land resources hall, and a temporary one are rebuilt. They vividly demonstrate

01

01.1958年在北京西四落成的地质博物馆大楼

对页
02. 外景
03. 平面
04. 北立面

01. The museum built in 1988, Xisi Beijing

Opposite page
02. Exterior
03. Layout plan
04. North elevation

02

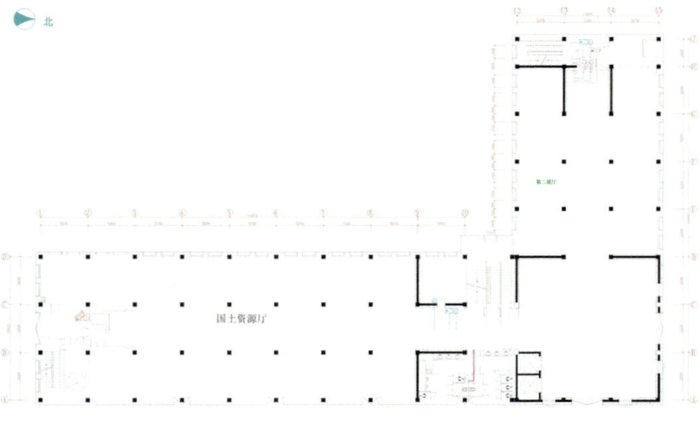

03

04

新的中国地质博物馆在设计理念和形式上均步入了世界领先行列。来华访问的英国博物馆代表团参观中国地质博物馆改造工程后，惊叹道："这是世界上最好的博物馆！"他们头脑里"中国博物馆落后陈旧"的观念被一扫而空。中国地质博物馆离"国内一流，国际著名"的目标已经越来越近。

national land resource, the earth science and beautiful specimen world.

The Geological Museum of China has reached international leading level in both design concept and form. The British museum delegation visited the renovation of the museum, marveled " This is the best museum !" The ideas of old museum backward in their head are now dispelled. The objective "Domestic first-class, international famous" of the Geological Museum of China has been approached.

05

06

05. 地球展厅
06. 矿物岩石展厅

对页
07. 宝玉石展厅
08. 史前生物展厅

05. Earth exhibition hall
06. Mineral rocks exhibition hall

Opposite page
07. Gems and jades exhibition hall
08. Prehistoric creatures exhibition hall

07

08

China Agricultural Museum

中国农业博物馆

中国农业博物馆是在全国农业展览馆的基础上成立的，于 1986 年 9 月正式向社会开放。中国农业博物馆的建筑颇具特色，环境十分优美，被誉为“中西合璧建筑思想在实践中应用的典范”。

中国农业博物馆是 20 世纪 50 年代北京十大建筑之一。馆区位于北京市东三环北路东侧，坐东向西与东直门遥遥相望。馆内总体布局分为 10 个单体建筑，各自有着不同的造型和特点。主体建筑一号馆高 33 m，正中有一座绿色琉璃瓦覆盖着的三重檐八角形楼阁，呈现出中国古典建筑特色。五号馆融入西洋古典建筑元素，正立面凸出三个高达 12 m 的三连圆拱门。七号馆最初为气象馆，圆形穹顶正中有一个高 2 m 饰有麦穗图案的宝顶，宝顶上竖立一个高 6 m 的风速风向仪。八、九、十号馆呈“品”字形排列，原为水产馆；其中号称“水晶宫”的十号馆小巧玲珑，最具特色。

China Agricultural Museum was established on the basis of National Agriculture Exhibition Center and was officially opened to the public in September 1986. China Agricultural Museum’s distinctive building and graceful environment make it be considered as a model of "Chinese and Western architectural thinking applied in practice ".

China Agricultural Museum is one of Top 10 Beijing Buildings in 1950s. The museum lies in the east of East Third Ring North Road, facing west across distant Dongzhimen. The overall layout of the exhibition hall is divided into 10 single buildings which have their own different shapes and characteristics. Hall 1 of the main building is 33 meters tall and there is a triple-eave octagonal pavilion covered with green glazed tiles in the center, showing a classical Chinese architecture feature. Hall 5 is blended in Western classical architecture elements and there are three protruding three-joined rounded arches of 12 meters in the front elevation. Hall 7 was is a meteorological hall originally. In the median of the domical vault is a 2-meter-high “Baoding” with pattern of ear and on the “Baoding” there is a six-meter-high anemoclinograph. Hall 8, Hall 9 and Hall 10 are arranged in the shape of a Chinese character “品”, originally called “Fisheries Museum”, among which, Hall 10, “Crystal Palace”, is the most famous for its small but exquisite features.

01

01. 一号馆雕塑

对页
02. 一号馆外景
03. 一号馆正立面
04. 一号馆平面

01. Sculpture of Hall 1

Opposite page
02. Exterior of Hall 1
03. Façade of Hall 1
04. Layout of Hall 1

02

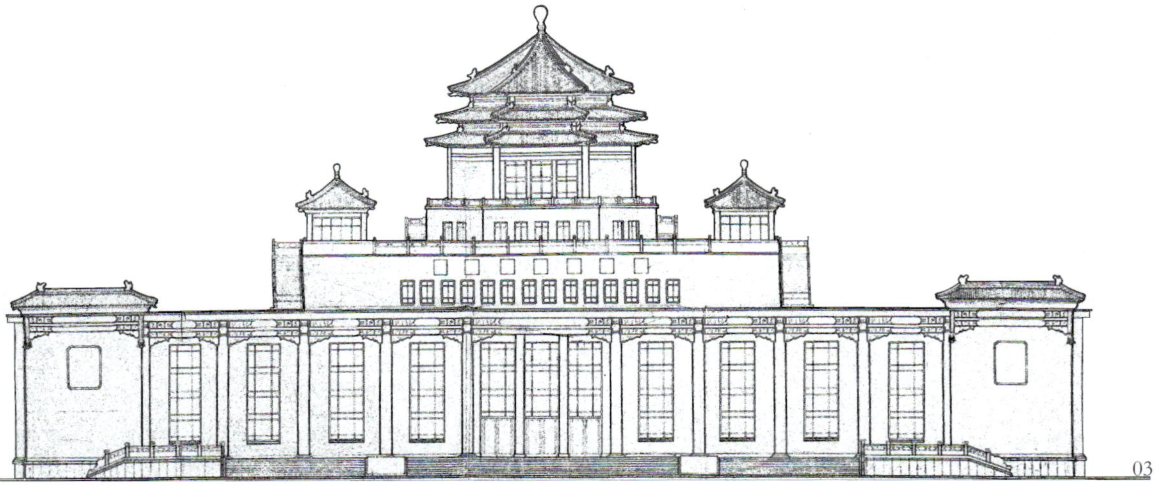

03

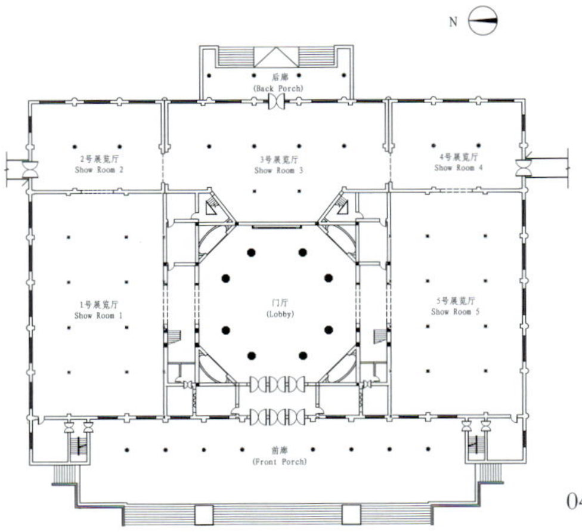
N
后廊
(Back Porch)
2号展览厅
Show Room 2
3号展览厅
Show Room 3
4号展览厅
Show Room 4
1号展览厅
Show Room 1
门厅
(Lobby)
5号展览厅
Show Room 5
前廊
(Front Porch)

04

05

06

07

建筑四周遍布苍松翠柏，绿树成荫，繁花似锦，与回廊楼阁、碧瓦朱檐交相辉映，体现了中国园林建筑中建筑与环境相协调的思想方法。建筑物上大量的农作物装饰图案——象征农作物茁壮成长的绿色琉璃瓦，寓意五谷丰登的米黄色外墙，与中心广场的喜庆丰收雕塑一起，从不同侧面突出了中国农业博物馆的农业特色。

Around the museum are all covered with pines and cypresses, which offer shade , with carpet of flowers, adding radiance to each other with pavilions and corridors, green tiles and red eaves, reflecting Chinese garden architecture thinking–harmonizing the building with the environment. On the building there are large numbers of decoration patterns of crops: green glazed tiles symbolizing the crops' growing, beige walls implicating bumper grain harvest and the central sculpture hinting the happy harvest, all of these designs highlight agricultural characteristics of China Agricultural Museum from many different aspects.

07. 园林廊亭秋色

对页
05. 七号馆外景
06. 水产馆外景

07. Landscape in the autumn

Opposite page
05. Exterior of Hall 7
06. Exterior of Aquaculture Hall

Chinese Aviation Museum

中国航空博物馆

中国航空博物馆是首批国家一级博物馆、爱国主义教育基地、首批全国国防教育示范基地和科普教育基地、国家 AAAA 级旅游景区，是军队向社会开放的“窗口”，国家向世界展示人民空军辉煌成就和航空国防科技事业迅猛发展的“窗口”，弘扬航天文化的课堂，承载空军精神的摇篮，宣传和谐航天的平台；是世界排名前五位、亚洲第一的集知识型、教育型、科技型、研究型、园林型、旅游型于一体的大型航空专业博物馆。

博物馆展览展示布局科学，建筑独特，含义深刻。有象征着 21 世纪独具空军特色的大门；有以飘扬的军旗为造型的胡锦涛主席题词墙；有代表空军武器组成种类和全馆展览内涵的馆标区；有折射空军文化、承载空军精神的英雄广场、英雄大道群雕和“利剑”主雕塑区；有独具匠心、气势宏伟的综合展馆和四季恒温、世界最大的洞库展厅等，令人惊叹。

Chinese Aviation Museum is in the first batch of state first-class museums, It is a patriotism education base, the first batch of national defense education and scientific education base, the AAAA scenic spot of the national level. It is the public channel the army opens to the society. It is also the public channel for the country to show the achievements of the air force and rapid development of air defense science and technology. It is the class to disseminate knowledge of aerial culture, the cradle to carry the air force spirits, the platform to propagate harmony. Chinese Aviation Museum is one of the world's top five museums , the first in Asia. It is a large-scale professional aviation museum which is a collection of knowledge, education, science and technology, research, garden and touring.

Chinese Aviation Museum with the unique building, reasonable layouts in exhibition and display, has profound implications. There is a distinctive gate of the air force which symbolizes the 21st century, a wall inscribed by President Hu which is casted as a waving banner, a symbolic zone which represents the air weapon types and the connotation of the entire exhibition. The Hero Square, the group of sculptures on the Hero avenue, the main sculpture area named "Lijian" all reflect air force culture and carry the air force spirits. Chinese Aviation Museum also has a comprehensive exhibition hall which is characteristic and imposing , and the world's largest cavity hall which can keep constant temperature in the four seasons and so on. They are all astonishing.

01

01. 雕塑

对页
02. 广场
03-04. 主展厅外景

01.Sculpture

Opposite page
02. Square
03-04. Exterior of the main exhibition hall

02

03

04

05

06

07

06. 前厅　　06. Entrance hall
07. 展厅　　07. Exhibition hall

对页　　**Opposite page**
05. 主展厅内景　　05.Interior of the main exhibition hall

China Railway Museum

中国铁道博物馆

中国铁道博物馆是中国铁路系统的国家级专业博物馆，现有三个馆：正阳门馆、东郊馆和詹天佑纪念馆，展出内容及陈列形式各具特色，建筑风格迥异。

正阳门馆是中国铁道博物馆的主馆，位于天安门广场东南侧，原为京奉铁路正阳门东车站旧址，占地面积 2 024 m^2，建筑面积 9 458 m^2，建筑高度 11 m，最高建筑（钟楼）为 21.2 m。车站整个建筑为典型的欧洲风格，1906 年落成，曾是早期中国铁路进出北京的重要门户。1999 年完成了复原改造工程，保留了原建筑的欧式浪漫典雅风格，最大限度地恢复了昔日车站的原貌；2008 年改建为博物馆，展区分为地下一层、地上三层，馆内陈列主要展现中国铁路发展的历史轨迹。

China Railway Museum is a national level specialized institution, which consists of 3 exhibition halls: Zhengyangmen Exhibition Hall, Eastern Suburb Exhibition Hall and Zhan Tianyou Memorial. They have a variety of display contents in different forms. Their architectural styles are also different from each other.

Zhengyangmen Exhibition Hall is the main exhibition hall of China Railway Museum. It is located to the southeast corner of Tian'anmen Square. As a restoration of the building of "Beijing Mukden Railway Zhengyangmen East Station", the museum covers an area of 2,024 square meters, and has a floor area of 9,458 square meters. The height of the main building is 11meters, and the top of station clock tower is 21.2 meters high. Taken into operation in 1906, the station used to be an important entrance accessing Beijing for early Chinese railway. After the completion of the restoration project in 1999, the station preserves its elegant European style and maximally resumes its original appearance to the maxinum, and in 2008 it has been transformed into a museum. The exhibition hall is devided into three floors upper ground and one floor under ground. It mainly presents the historical process of the development of China railway.

01

01. 正阳门馆细部

01. Detail of Zhengyangmen Exhibition Hall

02

02. 正阳门馆　　02. Zhengyangmen Exhibition Hall
03.1958 年的北京站　　03. Beijing railway station in 1958
04. 老货场　　04. Old freight yard

03

04

05

06

东郊馆位于北京市朝阳区酒仙桥北路 1 号院北侧，占地面积 61942 m^2，是一个大型展示场馆，主要包括机车车辆展厅、综合展厅和室外展示区。机车车辆展厅建筑风格现代、清新、简洁，主色调为蓝、白二色，建筑面积 16500 m^2。建筑形式为现代风格，以平面设计中的主轴线为中心，西部主要为展示空间，以实体的浮雕墙面为主，浮雕左侧三层错落重叠的片墙以不同的色彩和抽象的造型代表三代机车（蒸汽机车、内燃机车和电气机车）的发展变迁。

Eastern Suburb Hall is located in the north of Court No. 1 , Jiuxianqiao North Road, Chaoyang District. It covers an area of 61,942 square meters. The large-scale venue consists of a locomotive and car exhibition hall, a comprehensive exhibition hall and outside exhibition area. With a style of modern, simple and refreshing, the locomotive and car exhibition hall is mainly colored in white and blue, with a floor area of 16,500 square meters. The modern style architecture centers with the vertical line. Its west side is an exhibition space, the main part of which is a relief wall. In the left part of the wall, there are three overlapped layers in different colors and abstract shapes, representing the development and changes of three generations of locomotives (steam, diesel and electric locomotives).

05. 东郊馆全景
06. 东郊馆鸟瞰效果图

05. Panorama of Eastern Suburb Hall
06. Aerial view of Eastern Suburb Hall

07

詹天佑纪念馆坐落于举世闻名的八达岭风景区，是一座为纪念在我国近代铁路建设史上做出杰出贡献的爱国者詹天佑而建立的专题人物纪念馆。它建在京张铁路八达岭隧道最高点，顺依山势，造型古朴。建筑采用刚直的几何体和线与面的结合，构图简洁明快，空间构成丰富，序列布局多变，与蜿蜒曲折、富有韵律感的万里长城形成明显的对比。展厅大门入口旁的詹天佑大型浮雕，全长 41.8 m，高 5.4 m，中间以“之”字形铁路作为分割线，将浮雕分成红、灰两部分，布局构思巧妙新颖。

Located in the world-famous Badaling Scenic Area, Zhan Tianyou Memorial is a special character memorial. Zhan Tianyou Memorial is built in the highest point of the Badaling tunnel of Beijing-Zhangjiakou Railway. In contrast to the winding and rhythmic the Great Wall, built along the mountain, in a form of classic elegance, the architecture of the memorial adopts angular geometry and the combination of line and plane, which makes a simple, neat and changeful layout and a variety of space forms. Beside the entrance of the exhibition hall, composed in ingenious idea, the 41.8 meters long, 5.4 meters high, large-scale Zhan Tianyou relief is divided into two parts from the colors of red and gray. As a separating line, the middle of the two parts is the sketch of switchback lines.

08

07. 詹天佑纪念馆
08. 詹天佑大型浮雕

07. Zhan Tianyou Memorial
08. Large-scale relief of Zhan Tianyou

National Museum of Modern Chinese Literature

中国现代文学馆

中国现代文学馆新馆建设分为两期工程，一期工程已于 2000 年 5 月建成并正式对公众开放，二期工程目前正在建设过程中。

设计师在建筑设计中充分利用中国古典建筑优点的基础上进行了创新。首先，吸纳了古典建筑中坡形屋顶的风格，从远处望去非常美观，同时又有调节室内温度的实用价值。其次，与古典建筑不同的是，文学馆的建筑中屋顶的屋脊是被分开的，里面暗藏了一个玻璃采光带。这样一来房屋的跨度可以做得很大，室内的光线也更为充足，而室内屋顶的三角形端面造型也显得非常漂亮和美观。

另外，中国现代文学馆的整体布局并没有采用传统的正南正北的模式，而是有一个 45°的"偏位"。通过这样一个角度的变换，中国园林设计中经典的移步换景效果得以实现，来这里参观的游客从不同角度望去能看到不同景色。这样的整体设计使得文学馆充满了活泼的现代气息，吸引了众多的游客驻足观看。

这些设计中的创新性特点使得中国现代文学馆一期工程具有了既典雅又现代、既庄重又活泼的风格。中国著名建筑大师张开济先生看过新馆之后，兴奋地说："这是中国的现代建筑，具有突破性。"文学馆建筑设计的独特性由此可见一斑。

The construction of New National Museum of Modern Chinese Literature is divided into two phases. Phase I was accomplished and opened to the public in May, 2000. Phase Ⅱ is still in the course of construction.

The designer makes innovation based on the full use of the merits of Chinese classical architecture. Firstly, he accepts the style of slop roof, which is very beautiful from a distant view. Meanwhile, it has practical value of adjusting the indoor temperature. Secondly, he divides the ridge into two separated parts, between which there is a ribbon skylight of glass so that the span of the hall can be very long and the indoor light rays can be more abundant while the triangle end face of the roof is also very beautiful.

In addition, the pattern of the whole composition of the museum is not Chinese traditional with the directions of due south and due north. There is an angle of 45 degree inclined to one side, by which the effect of the design of Chinese garden can be realized–moving steps to change sights–from different angles, the tourists can see different scenes. Such a kind of whole design makes the museum full of vivid modern flavor and attracts a lot of tourists to stay to appreciate it.

These innovative characteristics give the first phase of the museum the style of elegance, modernity, seriousness and vividness. After visiting the museum, Mr. Zhang Kaiji, a master of Chinese architecture said excitingly, "It is Chinese modern architecture. It is a breakthrough." Thus the uniqueness of the architectural design of the museum can be seen in some degree.

01

01. 彩色玻璃隔栅

对页
02. 博物馆外景
03. 主入口

01. Stained glass tarrier

Opposite page
02. Exterior of the museum
03. Main entrance

02

03

04. 外景
05-06. 庭院

04. Exterior of the museum
05-06. Court yard

07. 东立面　07. East façade
08-09. 内景　08-09. Interior of the museum

China National Film Museum

中国电影博物馆

中国电影博物馆位于规划建设中的北京影视城内，建筑面积约38 000m²，是一个以中国电影艺术成就为主题，集博览、展示、交流、仪典、娱乐等功能为一体的国际化电影文化中心。

该馆由电影放映区、中国电影历史展览区、电影技术博览区以及其他辅助功能构成。设计方案以周围场地环境及电影艺术的特征作为设计的切入点，借助一系列大众化的通俗意象和非常规的设计元素，融合建筑与电影经历，将参观者带入一个特殊的体验。设计者以电影为主题，表达了对于艺术与娱乐、虚拟与现实以及大众文化与严肃作品等问题的思考。

为体现电影的艺术特征及纪念性建筑的特点，建筑本身呈现为一个单体的黑色方盒，内部空间则借鉴了电影蒙太奇与长镜头的手法，并将光参与到设计之中，以充分体现其戏剧性和场景化的环境。该项目是一个非常规的工程——非建筑形式的建筑，其设计理念超前，建筑形态复杂，特殊做法多。大量高大、无柱及异形空间，装饰

Covering an area of 38,000 square meters in the Beijing Studios, China National Film Museum is an international cultural center, which features the art of Chinese film by means of exhibition, display, exchange, official functions and recreation.

The museum contains theaters, an exhibition hall for the history of Chinese film making, an exhibition of film technologies, and other functional facilities. Through the use of a series of popular images and unusual architectural elements, the project has integrated the characteristics of film art with the physical surroundings, creating a unique experience for visitors. With its focus on film art, the structure makes visitors think about the relationship between art and entertainment, the fictional and the real, and popular and high culture.

In order to reflect the artistic features of the movie and the characteristics of the memorial buildings, the museum appears as a shape of single black box, whose interior space uses the montage and long take for reference, more over the lights is involved in the design and fully reflects its dramatic environment with scenes in it. The project is an extraordinary work, in other words, it's called a non-architecture form of building with an advanced design concept, complex architectural styles and lots of special technologies. Because of many tall , column-less and heterotypical spaces as well as the façade whose decorative surface is separated from the maintenance side, architects have to break the conventions and adopt

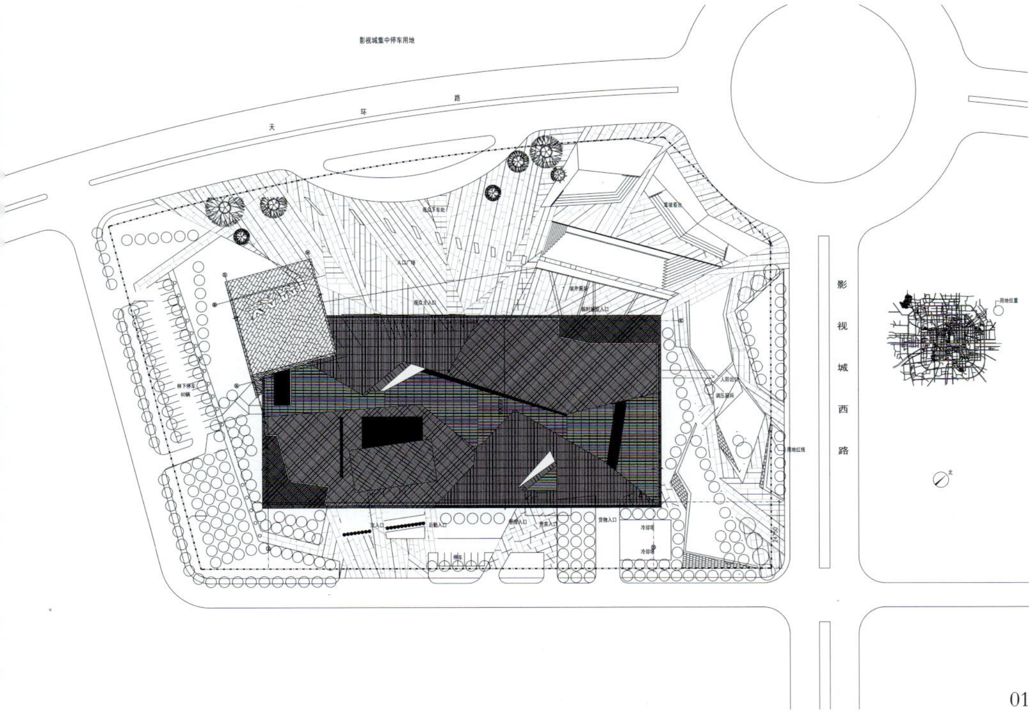

01

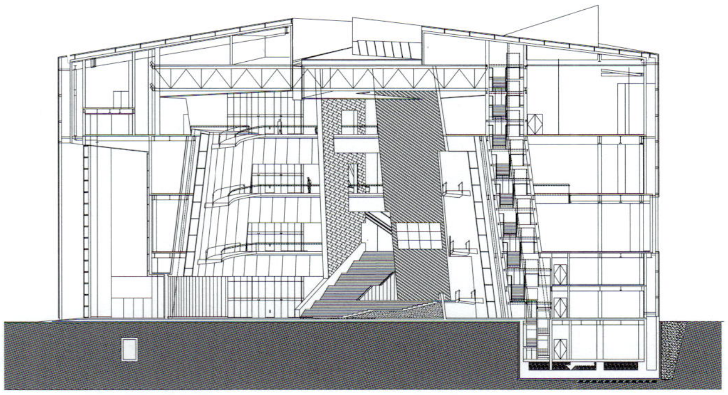

02

01. 总平面　01. General layout plan
02. 剖面　02. Section
03. 夜景　03. Night view

03

面与围护面分离的外立面设计，迫使设计突破常规做法，采用三维空间的设计手法。在幕墙设计、室内设计、光环境设计、LED 照明设计、景观设计等方面均有创造性突破，达到国外同类建筑的先进水平。

该工程是配合 2005 年的中国电影百年庆典活动的配套项目，同时也作为 2008 年北京奥运会的配套项目。

the design of three dimensional space. The building reaches the international advanced level and embodies lots of creative breakthrough, including the curtain wall design, interior design, illumination environment design, LED lighting design and landscape design.

The building was designed for the 2005 Art Festival of the Centennial Celebration of Chinese Films as well as for the 2008 Olympic Games.

05

04. 细部
05. 主入口
06. 主入口前灰空间

04. Detail
05. Main entrance of the museum
06. Grey space in front of the main entrance

04

06

07

08

07. 入口大厅
08. 南入口大厅

07. Hall of the main entrance
08. Hall of the south entrance

09

10

10. 中央圆厅

对页
09. 走廊

10.Central round hall

Opposite page
09. Corridor

Capital Museum, China

首都博物馆

首都博物馆是一座拥有先进设施的现代化综合性博物馆，是北京市政府投资兴建的面向 21 世纪的大型现代化文化设施，位于西长安街延长线白云路上。该建筑以其独到的建筑设计理念和优秀的建筑质量，荣获 2006 年度中国建筑工程鲁班奖，并在 2009 年被评为“北京当代十大建筑”之一。

首都博物馆建设项目的立项申请，于 2000 年得到市政府批准，2001 年经国家计委批准实施，2001 年 12 月 25 日正式奠基兴建，2005 年 12 月 15 日开始试运行，2006 年 5 月 18 日正式开馆。

首都博物馆的建筑设计理念是“以人为本，以文物为本，为社会服务”，强调“过去与未来、历史与现代、艺术与自然的和谐统一”。倾斜的青铜体破墙而出，衍生出文物发掘的意象；悬挑的大屋顶寓意中国传统建筑的出檐；木质幕墙传承了中国传统建筑的木结构；干挂石墙完美地链接了古代城墙与现代幕墙；广场的起坡取材于宫殿台基的手法，烘托了宏伟的建筑。

The Capital Museum is a modern comprehensive museum with advanced facilities. As a major modern cultural construction project facing to the 21th century, the new Capital Museum finally makes its debut on Baiyun Road, along the western extension of Chang'an Street. With its magnificent architectural conception and excellent quality, the museum won the national Lu Ban Prize in 2006 and was selected as one of "the best 10 modern architectures in Beijing" in 2009.

The new Capital Museum, with its application for construction approved by Beijing Municipal Government in 2000 and further approved by the SDPC (State Development Planning Commission) in 2001, finally commenced its construction on Dec. 25th, 2001, and then made its test-run on Dec. 15th, 2005 and the official business on May 18th, 2006.

The concept of the architectural design of the Capital Museum is based on human and cultural heritage to serve the community and underlines the harmonious integration of past and present, history and modernism, art and nature. The oval-shaped Bronze Exhibition Hall implies the unearthing of ancient relics by its slanting design in which extends from the ground to the exterior of the museum. The overhanging massive roof symbolizes the cornice of the Chinese traditional architecture. The timber curtain wall inherits the wooden structures in the traditional architecture. The application of the dry-hang slab accomplishes a perfect annexation of the ancient circumvallation and the modern curtain wall. The gradient at the entrance square is originated from the platform of the old palaces and increases the grandiosity of the building.

01. 日景
02. 外景
03. 总平面

01. Day view of the museum
02. Exterior of the building
03. General layout plan

01

02

03

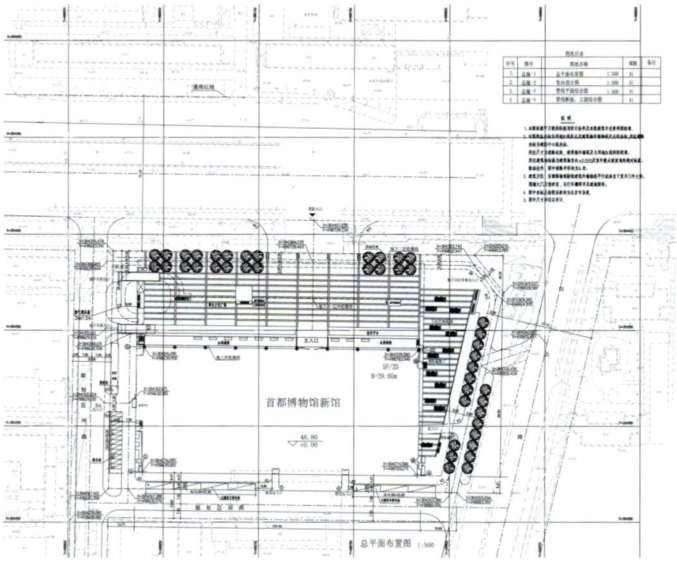

04

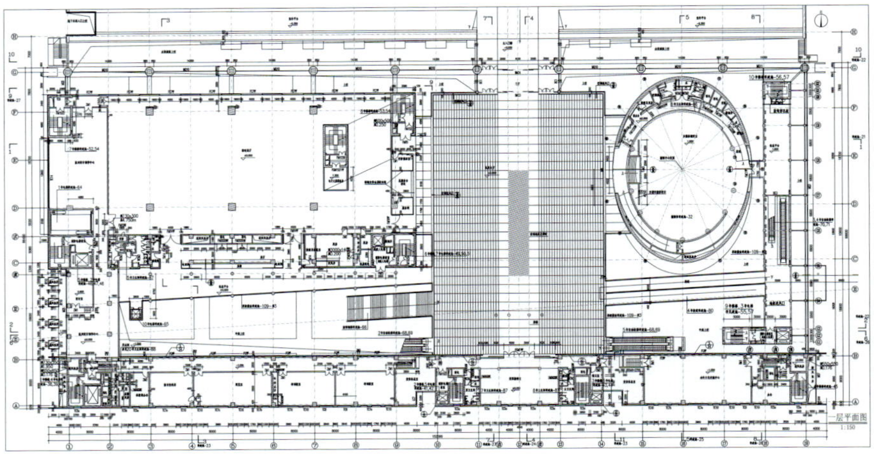

05

04. 首层平面　04. Ground level floor plan
05-06. 剖面　05-06. Section

对页　**Opposite page**
07-08. 大堂　07-08. Main foyer
09. 大堂青铜装饰　09. Bronze decoration in the main foyer

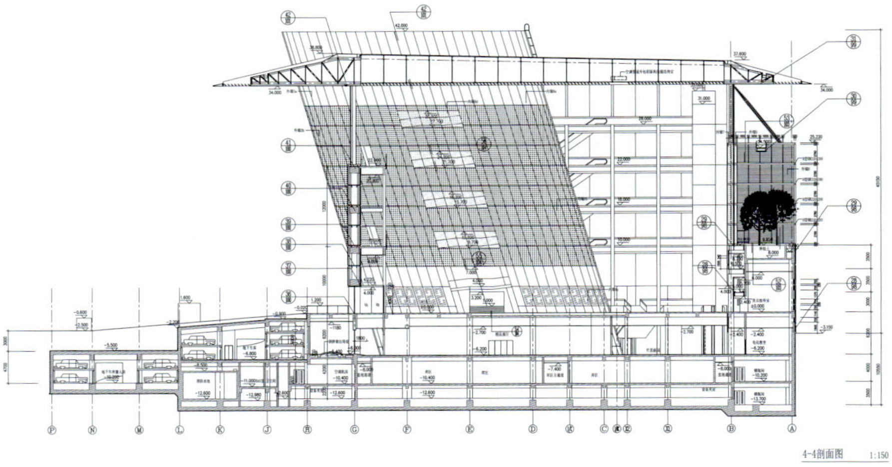

06

首都博物馆的建筑设计特点是以简洁的矩形平面与北京城市格局相协调，而非对称的形体则呼应街角空间；青铜、木材、石墙等传统建筑材料代表老北京的历史，先进的建造技术体现现代化的北京，光的反射和映象表达了历史与未来的对话；打破传统博物馆空间封闭、沉闷的感觉，为市民营造了一个开放、温馨、明亮的文化休闲环境。

首都博物馆主要的功能区分为展陈区、社会教育区、综合服务区、业务科研区、行政办公区、安全保卫区、藏品库区、地下停车库、设备机房区。整体建筑功能分区合理。

The sententiousness of the rectangular planning is in accordance with the city planning of Beijing. The dissymmetrical masses well responses the corner space. A large quantity of bronze, timber, stones and other traditional materials are used in the museum to present the history of amcient Beijing city. The advanced constructional technique incarnates the modernization of Beijing. The breaks and reflection from light expresses the dialogue between history and future.The new museum breaks the closed and depressive feeling from the traditional museums and builds the citizens an open, warm and bright environment for cultural and leisure activities.

The Capital Museum is divided into several functional spaces for exhibition halls, social education halls, comprehensive service halls, business and academic research halls, administration and offices halls, public security halls, store depots, underground parking and the facility rooms. All these functional spaces are arranged into a rational system.

奥运特许商品零售店
Olympic Merchandise Store
临时展厅
Temporary Exhibition Hall
工艺品店
Craftwork Store
书店
Book Store

07

08

09

11

12

13

14

15

16

11-14. 展陈区
15. 社会教育区　16. 综合服务区

对页
10. 大堂

11-14.Exhibition halls
15.Social education area　16. Integrated service area

Opposite page
10.Main foyer

Beijing Museum of Natural History

北京自然博物馆

北京自然博物馆是新中国依靠自己的力量筹建的第一座大型自然科学类博物馆。主要从事古生物、现代生物和人类学的标本收藏、科学研究和科学普及工作。现在博物馆馆藏标本数量、质量在国内名列前茅，科学研究和陈列展览水平在国内也均处于领先地位，是中宣部和北京市政府命名的"全国青少年科技教育基地"、"北京市爱国主义教育基地"，并于2008年被国家文物局评为国家一级博物馆。目前，本馆占地面积约14 072m²，展厅面积约10 000m²。

北京自然博物馆的前身是成立于1951年4月的中央自然博物馆筹备处。建馆初期，中央自然博物馆筹备处使用故宫博物院东华门内文华殿、传心殿、清史馆等办公和展出，房屋总面积5 137m²。

北京自然博物馆现坐落于天桥南大街，南中轴路东侧，南北长112m，东西宽61.2m。整个大楼由两个展览楼和一个标本楼共同组成，其中主楼整体为两层，高15.5m，局部三层，最高处24m，始建于1958年，当时中苏交流频繁，而受政治

Beijing Museum of Natural History is the first large-scale natural history museum built on our own strength after the foundation of People's Republic of China. The museum bears three main functions: collecting specimens, conducting academic research and popularizing scientific knowledge in the field of paleontology, biology, and anthropology. Endowed with the quality and numerous collection items and leading scientific research and exhibition design competency, Beijing Museum of Natural History has been named as"National Youth Science and Technology Education Base" and "Beijing Municipal Patriotism Education Base" and, in 2008, recognized as a national top-class museum by State Administration of Cultural Heritage of China.

Beijing Museum of Natural History grew out of the preparation department of National Central Museum of Natural History founded in April 1951. During the early period, the preparation department of National Central Museum of Natural History made its office and galleries inside the East Prosperity Gate (Donghua Men) of the Palace Museum in the Hall of Literary Glory (Wenhua Dian), Hall of Sacrificial Rites (Chuanxin Dian) and Hall of Dynastic History (Qingshi Guan), totally taking up 5,137 square meters.

Located in Tianqiao South Street, east to the South Central Road, Beijing Museum of Natural History covers a 112-meter-long and 61.2-meter-wide area. The main edifice encompasses two exhibition buildings and a collection building, the major block being two stories, 15.5 meters high and some sections

01

01. 博物馆旧影

对页
02. 博物馆外景
03. 田家炳标本楼

01.Former exterior of the museum

Opposite page
02. Exterior of the museum
03. Tian Jiabing collection building

02

P70-71 页
04-05. 动物新厅全景
06-07. 恐龙新厅全景

P70-71
04-05. Panorama of the new Zoological Hall
06-07. Panorama of the new Dinosaur Hall

因素影响，北京市建筑设计研究院的建筑师们在设计时，把传统的中国建筑元素与苏联建筑风格完美地结合在了一起。后于 1992 年建造了标本楼，并于 2002 年再建了一个展览楼，建筑规模由最初的 9 590.5m^2 扩至 18 533m^2；展览面积由原来的 5 865m^2 增至 9 988m^2，占了总建筑规模的 54%；后两次的扩建在建筑外观上均与 1958 年的展览楼保持了原有建筑风格的统一。

being three stories, 24 meters at the highest. The first building was started in 1958, and Beijing Institute of Architectural Design integrated traditional Chinese architectural elements with Russian architectural style as a result of political influence for China and the Soviet Union made frequent exchanges then. Later in 1992, the collection building was built and in 2002 the second exhibition building was built, expanding the architectural scale from 9,590.5 square meters to 18,533 square meters. The exhibition area was increased from 5,865 square meters to 9,988 square meters, taking 54% of the total architectural scale. The external appearance of the two later expansion projects maintained unity in architectural style with the 1958's exhibition building.

03

04

05

生命向陆地发展
FROM WATER TO LAND

06

EVOLUTION OF PROBOSCIDEANS
长鼻类的演化
安全出口
EXIT

07

什么是鱼
鱼类
FISHES

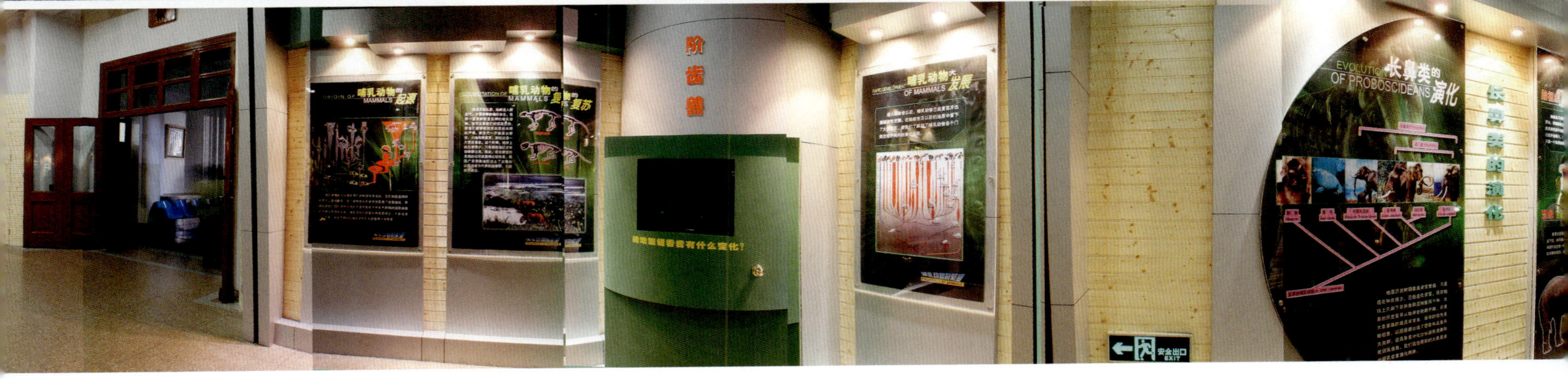
EVOLUTION OF PROBOSCIDEANS 长鼻类的演化
安全出口
EXIT

Beijing Planetarium

北京天文馆

北京天文馆（老馆）是中国第一座天文馆，位于北京西直门外，于 1957 年 9 月建立，建筑面积 4 337 m²，砖混结构。

展馆直径为 23.5 m 的天象厅象征天穹，中间安装精致的大型天象仪，可表演日月星辰运转和流星、彗星、日食、月食等天象，能容 400 人观看。门厅正中有反映地球自转的傅科摆。西侧展厅有天文知识展览，东侧展厅经常举行学术交流。庭院中有一座口径 13 cm 的望远镜，观众通过它观看月亮、行星、星云、星团，白天观测太阳黑子。

北京天文馆新馆于 2001 年底在天文馆原址上动工兴建，总建筑规模 21 000 m²，于 2004 年 12 月建成开放。

馆内主要的公共开放设施有：数字化宇宙剧场、3D 动感天文演示剧场、4D 动感影院、天文展厅、太阳观测台、大众天文台、天文教室等。宇宙剧场有 200 座席，引进国外先进的数字化天文放映设备，能生动形象地演绎壮丽的星空景象和人类探测太空的壮举。3D 和 4D 动感天文剧场，

Beijing Planetarium (Building A) is the first planetarium in China, located in Beijing Xizhimenwai, in September 1957 to establish a brick-concrete structure building area of 4,337 square meters.

A large-scale sophisticated planetarium is installed in the middle of the space theater, which is 23.5 meters in diameter and symbolizes the heaven. The theater accommodates 400 people watching the movements of the moon, stars, meteors and comets, the solar and lunar eclipses and other celestial phenomena. There is a Foucault pendulum which reflects the Earth's rotation in the middle of the entrance hall. Exhibitions about astronomical knowledge are held in the west hall and regular conferences are held in the east one. Inside the courtyard stands a 13-centimeter telescope. By using it the visitors can get observations of the moon, stars, nebulas and star clusters, even maculae in the daytime.

Building B started its construction on the same site of the building A, total area of 21,000 square meters, finally completed and opened in December, 2004.

The main public facilities in the planetarium include digital space theater, 3D dynamic theater for astronomical shows, 4D dynamic theater, astronomic exhibition halls, sun observatories, public observatories and astronomic classrooms. Using the advanced digital projection machine imported from overseas, the 200-seated theater can display the

01

02

对页

01. 天文馆旧影

02. 北京天文馆 B 馆入口

03. 外景

04. 顶部全景

05. 新馆太阳观测台近景

Opposite page

01. Old photograph of Beijing Planetarium

02. Entrance of Beijing Planetarium (Building B)

03. Exterior

04. Panorama of the top-telescope

05. Close view of Building B observatories

06

06. 侧照
07. 新馆内部一角

对页
08. 让人身临其境的 4D 剧场
09. 动感十足的 3D 剧场
10. 数字宇宙剧场内部
11. 光学剧场内部
12. 绘有古代天文故事的天花板

06. Side view
07. A corner inside Beijing B

Opposite page
08. 4D Dynamic Theatre
09. 3D Dynamic Theatre
10. Interior of Digital Space Theatre
11. Interior of Optical theatre
12. Ceiling painted ancient astronomical stories

07

其容量分别为 48 席和 196 席座位，以其强烈的动感、特殊的视听合成效果为观众营造出很强的临场感。太阳观测台展览面积 300 m^2，以展出太阳和太阳系内容为主。其主要展示内容有从太阳真空望远镜观测接收而来的 1.8 m 太阳白光投影像，174 mm 手描太阳黑子白光全日面投影像，太阳光谱投影像，太阳白光、太阳色球全日面和太阳局部活动区的 CCD 电视图像等。

splendid perspective of the universe and the great feats of human beings to explore the outer space dramatically. The 3D and 4D dynamic theater have 48 and 196 seats and show the visitors a strong spontaneous experience with the fierce dynamic visual and audio effects. The sun observatory is 300 square meters large and displays the sun and the solar system mostly. The major contents include the 1.8 meters sun white projection images from the solar vacuum telescope, the 174-mellimeters hand-painted white full-face sunspot projection images, solar spectrum projection images, white sun, solar monochromatic images and the CCD television images of the active region on the sun.

08

09

10

11

12

Zhoukoudian Relics Museum

周口店北京人遗址博物馆

周口店北京人遗址博物馆建于1971年，建筑面积1 036.32 m²。“回”字形结构，博物馆中部为天井，起到局部采光的作用。

周口店北京人遗址博物馆新馆占地面积38 072 m²，建筑面积7 989 m²，建筑限高9 m，分地下一层和地上二层，投资约1.92亿元。周口店北京人遗址博物馆新馆建筑外形创意源于“北京人”50万年前制作和使用的劳动工具——石器。建筑线条简约、粗厉、刚毅，多折面的面与面之间形成锋利的石器“刃口”结构。建筑由南向北逐渐降低，高度的变化避免了建筑形体的呆板，与周边山脉、天际线及自然环境巧妙地融为一体。

博物馆新馆将集陈列展览、科普教育、藏品保管、科学研究和管理办公于一体，将在展览形式、展陈内容、服务标准、管理功能上展示出国际水准。主体建筑计划于2011年完成。

Zhoukoudian Relics Museum was built in 1971, and the floor area is 1,036.32 square meters. For local lighting, the central part of the museum is courtyard, just like the Chinese character“回”.

New museum covers an area of 38,072 square meters, and floor area is 7,989 suqare meters, limit for height of the architecture is 9 meters, costing 192 million RMB, consisting of two floors on the ground and one floor underground. Architectural appearance originality for the museum comes from stone tools of Peking Man. The architecture line is simplicity and fortitude. Construction drops down from south to north, skyline and the natural environment are harmoniously integrated.

The new museum will focus on display, popular science education, collections storage, research and management as a whole. It will show an international level on display, service standard, and management function. It is expected to be finished in 2011.

01

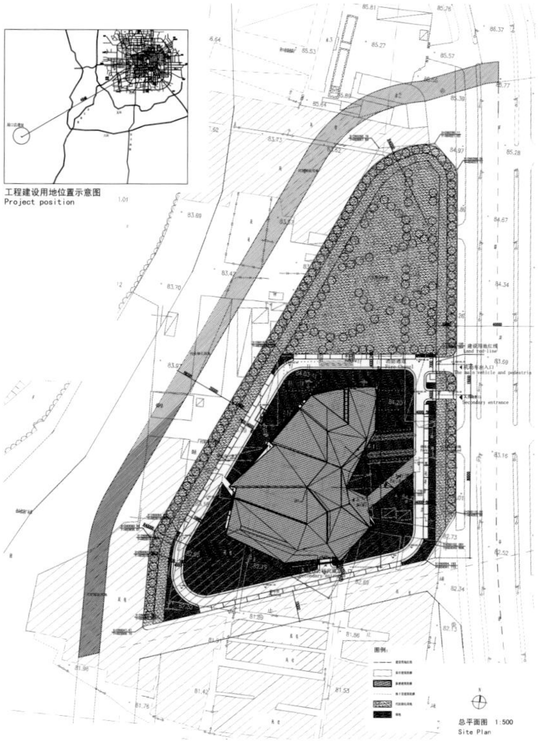

02

02. 博物馆新馆总平面
03. 博物馆新馆首层平面
04. 博物馆新馆东立面
05. 博物馆新馆西立面
06. 博物馆新馆效果图

对页

01. 博物馆旧馆鸟瞰

02. Master plan
03. 1st floor plan
04. East elevation
05. West Elevation
06. Rendering

Opposite page

01.Aerial view of the museum

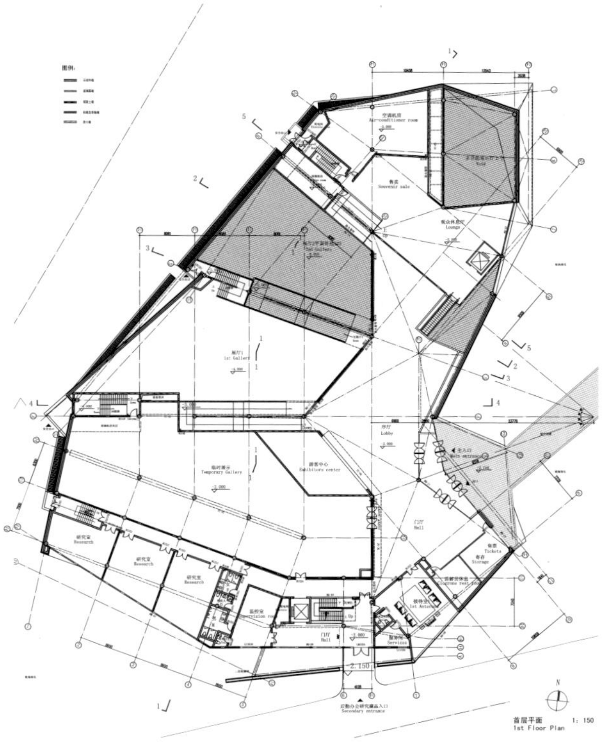

03

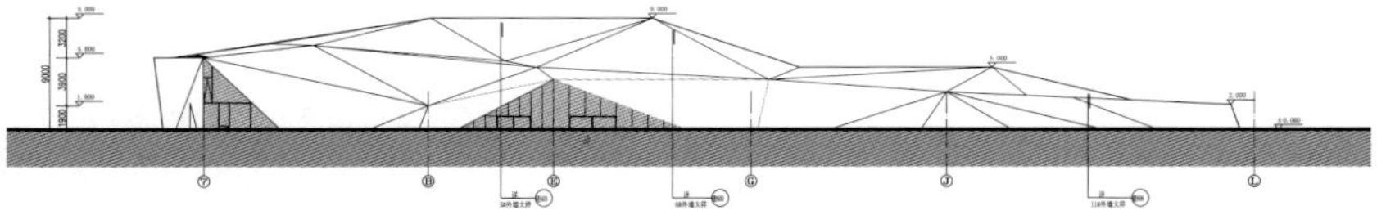

04

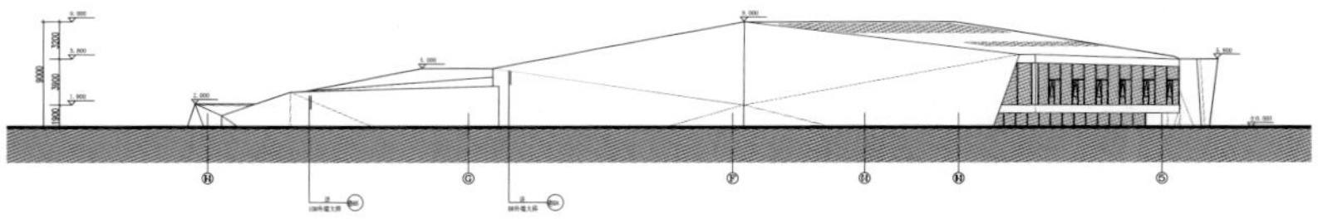

05

06

Fangshan Global Geological Park

房山世界地质公园博物馆

房山世界地质公园涵盖北京房山区及河北省涞水县、涞源县的相关区域。该馆以其丰富的地质资源，优美的自然环境，优越的地理位置和多样的生物类型，于2006年获得联合国教科文组织授予的“世界地质公园”桂冠，使北京成为世界上第一个拥有“世界地质公园”的首都城市。房山世界地质公园也是华北第一家世界地质公园，而房山世界地质公园博物馆为地质公园揭碑开园的重点工程，不仅起到科普展示、典型地质构造演示的作用，同时作为房山区的地标性建筑起到区域名片的作用。

本项目位于北京市房山区长沟镇西北路，建设用地47700 m²，位于至京西古刹云居寺的必经之路南侧，总建筑面积10000 m²。其中博物馆面积5000 m²，包括展厅及4D影院、旅游咨询服务及培训等配套设施，主体建筑3层，总高24 m。

建筑师以“设计编译自然”为理念，通过对自然环境的解读，梳理建筑与山脉水体等环境要素的关系；用人工建筑修复自

Fangshan Global Geological Park covers a large area in Beijing Fangshan District, Laishui County and Laiyuan County of Hebei Province. With its rich geological resources, beautiful environment, excellent location and biological diversity, it was awarded the honor of UNESCO "World Geological Park" in 2006, which makes Beijing the world's first capital city to ever earn a title like this and makes itself the first geological park in the North China. The ceremony for the opening of the Fangshan Global Geological Park, which proposed by the museum, not only has the function of science exhibition and typical geological structure presentation, but also has been the landmark of the whole Fangshan District.

The project is located in Xibei Road, Changgou Town of Fangshan District and covers an area of 47,700 square meters. It sets in south of the only way to the Yunjusi Temple, the famous ancient architecture in the west of Beijing, with a total floor area of 10,000 square meters, half of which is the museum, including the exhibition halls and a 4D screening hall; the rest buildings are mainly for travel advisory services, training and other facilities. The main building of three floors is totally 24 meters hight.

Holding the concept of "Design to compile the nature", the architect refined the relationship between buildings and mountains, water and other environmental elements through the interpretation of the nature. Artificial constructions are made to combine with the original hills to make the whole complex a group of rock mountains, thus with the migration of

01

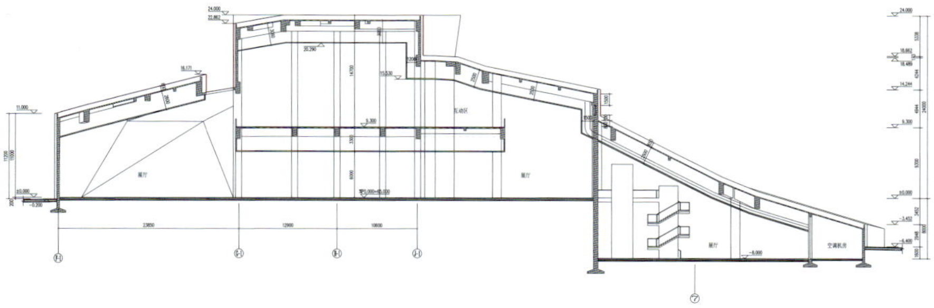

02

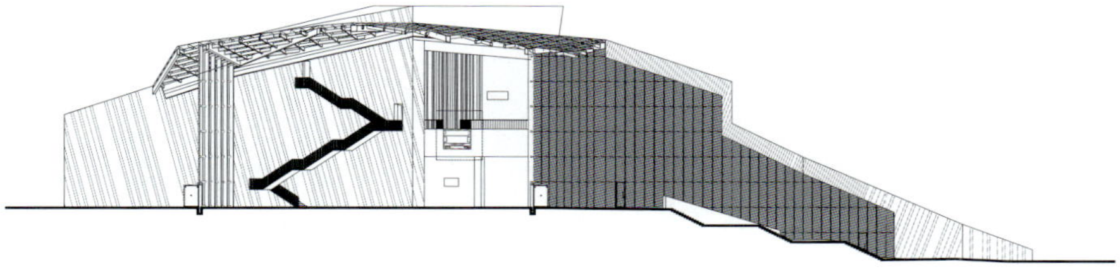

03

01. 外景　　01. Exterior
02. 剖面　　02. Section
03. 立面　　03. Elevation
04. 北立面细部　　04. Detail of the north elevation

04

05

06

然，形成如岩石山体般硬朗的整体造型；虚实相生的三大体量的互相融合，形成各功能空间的独立与流通。在建筑表皮的选用上，以房山地区特有的绿色板岩包裹整个建筑，使整个建筑宛如从地表隆起的巨石，与一岱远山遥相呼应，凸显了本土建筑文脉的主题，弘扬了建筑与场所的整体精神，并表现建筑师对自然的无比崇敬之情。

spaces, it has both the independence and mobility in different room. In the selection of the surface, a unique type of green slate in the district was chosen to cover the whole architecture which makes it a gigantic rock rising from underground with a mountain echoed. It highlights the theme of the local architectural context; promotes the spirit of the overall place, and shows the incomparable performance of the architect to the nature.

07

08

07. 二层室内
08. 室内

对页
05. 主入口
06. 主入口南侧

07. Interior of the second floor
08. Interior

Opposite page
05. Main entrance
06. South of the main entrance

Tianjin Museum

天津博物馆

天津博物馆坐落于天津市河西区友谊路与平江道交口的银河广场旁，是天津市重要的文化标志性建筑。

天津博物馆占地面积 57 000 m²，总建筑面积 35 032 m²，拥有 11 000 m² 的现代化展厅和功能齐全的文化休闲设施。建筑共三层，最高处 33 m。主体部分结构形式为钢筋混凝土框架—剪力墙钢支撑结构，屋顶为钢网壳，建筑平面呈半圆形，直径达 186 m，基础形式为桩—承台基础。

设计师运用现代浪漫主义的思维观点，引用仿生原理，以极具震撼力的创作手法，造就了一座昂首挺立、展翅腾飞的“白天鹅”仿生造型建筑，赋予城市景观以生命活力，从而使天津博物馆的外观与周围优美环境和谐地融为一体，极富时代感与艺术性，颇具特色。

天津博物馆在建造过程中，采用了 12 项专利技术，包括可“呼吸”的大跨度空间结构、清水混凝土施工工艺、镀瓷铝板屋面、弦支穹顶网壳结构、保温螺栓、板铰

Tianjin Museum is located at Yinhe Square, where is the cross of Youyi Road and Pingjiang Road in Hexi District. It is the important cultural landmark of Tianjin.

Tianjin Museum covers an area of 57,000 square meters, including floor area of 35,032 square meters and exhibition area of 11,000 square meters. The modern exhibition hall is equipped with multi-functional cultural and recreational facilities. The museum is a three-storey building, and reaches 33 meters at its highest point. The building, with 186-meter diameter, takes the pile-cap system as the foundation, the steel-shear-wall structure as the main body frame, and the steel grid shell as the roof. The plane view of the building is completely semicircular.

The designer applied a romantic and bionic idea to the architectural style, working out the appearance of the building as a flying swan. It symbolizes the rapid development of Tianjin and makes the city full of youth and vitality.

The designer applied 12 proprietary technologies to the architectural design, including large-span spatial structure, chord bracing system, waterproof domed roof, insulating bolt, zincification technology, hexahedral sloping wall, and

01

02

节点体系、锌加技术、大直径高强螺栓、大拱及铸刚节点、六面斜墙、屋面防水设计、超长度异型压型钢板，使得整个建筑具有超前性。

天津博物馆的整体造型为半球形结构，这种结构并不十分利于发挥博物馆的功能性，因此，在内部深化设计和陈列布展时，因地制宜，根据文物展陈特点和职工办公需求，把结构劣势转化为功能优势，收到了意想不到的良好效果。

天津博物馆开馆至今，无论建筑外观还是展览内容都得到了广泛认可，真正成为美观与实用性、建筑新工艺与人性化服务相结合的完美的建筑形式。

so on,which made the building ahead of this time.

The administration authorities applied human-oriented design to the exhibition arrangement, to develop the function superiority of the architecture, and to achieve a better balance between the museum and the audiences.

Tianjin Museum has always enjoyed a high reputation in architectural appearance and exhibition contents since it opened. It's a perfect combination of beauty and practicality, of new techniques and humanistic services.

03

02. 博物馆外景 02. Exterior
03. 细部 03.Detial

对页 **Opposite page**
01. 室内 01. Interior

04

05

04-06. 室内

04-06. Interior

Hebei Museum

河北省博物馆

河北省博物馆是河北省唯一的省级综合性博物馆，1953 年 4 月始建于当时的河北省会保定市。初建时的河北省博物馆位于保定市古莲池西南角的松鹤园内，是一座二层灰砖小楼，面积约 1 200 m^2，有 4 个展厅。

1981 年，河北省博物馆迁至河北省的新省会石家庄市，1986 年 12 月与原省展览馆合并，迁入现址。河北省博物馆现馆舍建成于 1969 年 12 月，是当时的“毛泽东思想胜利万岁展览馆”。馆舍坐落在石家庄市中心地段，面临宽阔的省会文化广场，总建筑面积 20 028 m^2，展览面积 11 216 m^2，有 18 个展厅。

主楼为仿人民大会堂风格的廊柱式建筑，既庄严朴素，又巍峨壮丽。主楼正面东西两侧墙壁分别有大型锻铜壁雕——“燕赵古韵”和“太行雄魂”，正门两侧有大型青铜雕塑“双翼神兽”。建筑内部的各式顶灯、吊灯和壁灯均设计为葵花的形状，典雅大方。2001 年，该建筑被公布为河北省第四批文物保护单位。

Hebei Museum is the only comprehensive museum of Hebei Province. It was founded in Baoding City in April 1953 which was the capital city of Hebei Province at that time. It, a two-storey brick dust building, was first located in the southwest corner of the Ancient Lotus Pond Garden. With a total floor area of 1,200 square meters, it contained 4 exhibiting halls.

In 1981, Hebei Museum removed to Shijiazhuang, the new capital city of Hebei Province, and moved to its current location after amalgamating with former Hebei Exhibition Hall in December 1986. As the Exhibition Hall of Long Live Thoughts of Chairman Mao, the building of Hebei Museum was completed in December 1969. Facing the Culture Square of Shijiazhuang, it is located in the city center. Its floor area is approximate 20,028 square meters, while the exhibition area occupies 11,216 square meters and is divided into 18 exhibition galleries.

Imitating the porch architectural style of the Great Hall of the People, the main building is simple, elegant and magnificent. The east and west flanks of building's facade are decorated with brass sculptures, "Yanzhao Ancient Charm" and "Taihang Heroic Spirit". A couple of bronze two-wing mythical beasts is situated on both sides of the main portal of the museum. With elegance and taste, all of the lights inside the museum were designed in the shape of sunflowers. The building was appraised as the Important Protection Unit of Cultural Relic of Hebei Province in 2001.

01

01. 博物馆旧影（摄于 1969 年，时为“毛泽东思想胜利万岁展览馆”）

对页
02. 博物馆鸟瞰
03. 博物馆南立面

01. Historical Scene of the Museum (In 1969, the museum was named "The Exhibition Hall of Long Live Thoughts of Chairman Mao" at that time.)

Opposite page
02. Aerial view of the museum
03. South facade of the museum

02

03

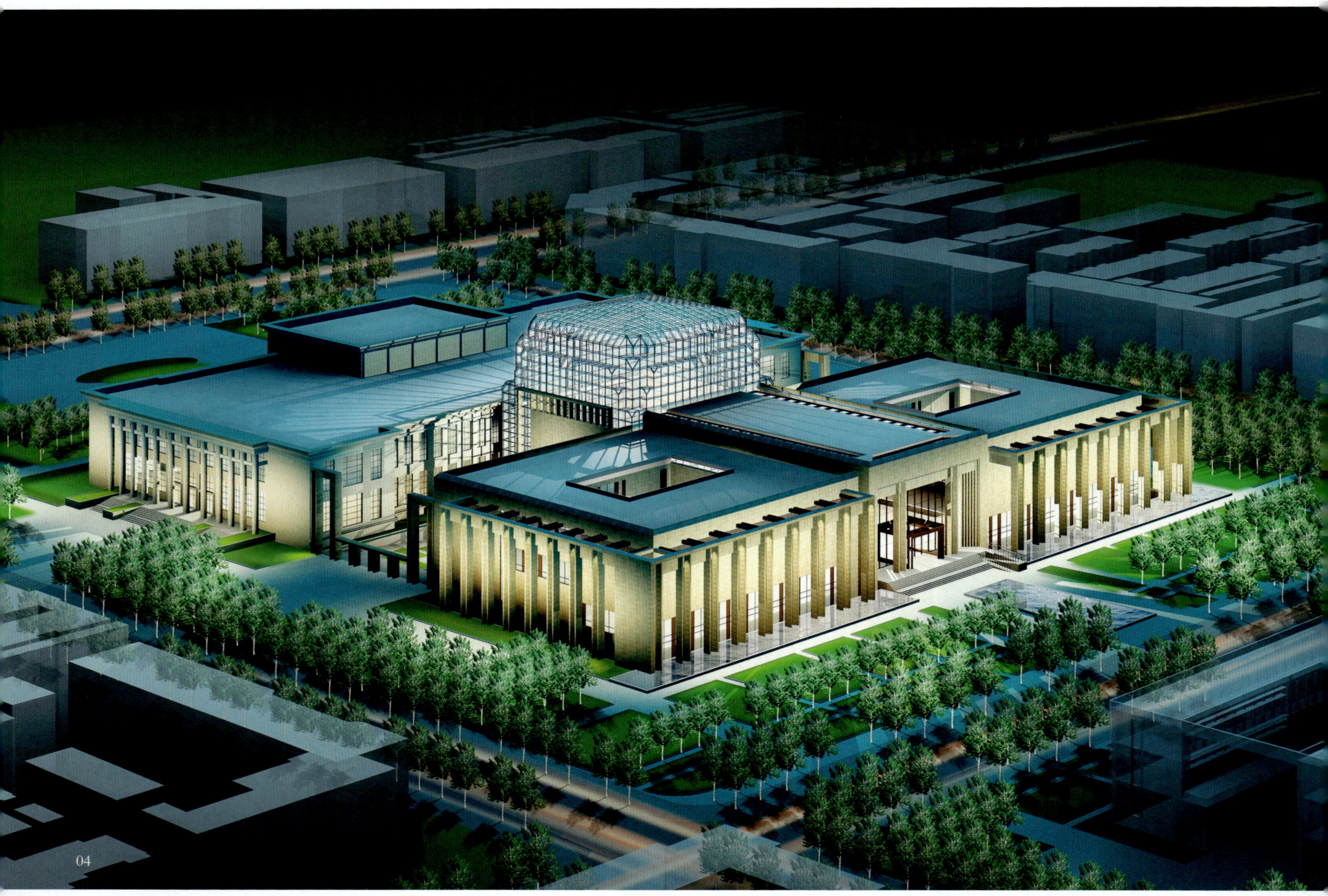

目前河北省博物馆正在扩建，新建馆舍位于现馆址的南侧，于 2006 年 9 月 6 日奠基，现正在加紧建设，总建筑面积 30 000 m²，展览面积 11 500 m²。新馆按照外观新颖、内部实用、整体和谐、体现特色的思路设计，与旧馆风格协调、相得益彰。

On the south side of the old museum, a new museum project has laid a foundation in September 2006 which is under construction. The new one has a floor area of 30,000 square meters, while the exhibition occupies 11,500 squane meters. New building in accordance with the design ideas of novel appearance, internal utility, harmonious integration, reflecting the characteristics, harmonizes with the old building in color and style, which makes them reinforce each other.

04. 新馆鸟瞰效果图

对页
05. 新馆西南侧透视效果图
06. 新馆主入口效果图
07. 新馆内庭效果图

04. Perspective of the new museum

Opposite page
05. Southwest perspective of the new museum
06. Perspective of the entrance of the new museum
07. Internal courtyard of the new museum

05

06

07

Xibaipo Memorial

西柏坡纪念馆

西柏坡是河北省平山县一座风光秀美的小山村，1948 年 5 月至 1949 年 3 月期间，中共中央曾驻在这里，并在这里指挥了闻名中外的辽沈、淮海、平津三大战役，被称为“解放全中国的最后一个农村指挥所”。1949 年 3 月中共七届二中全会在这里召开，会后不久中共中央离开西柏坡迁往北京，完成了建立新中国的伟大征程，被誉为“新中国从这里走来”。

西柏坡纪念馆馆区位于西柏坡村东侧，背依松柏葱翠的柏坡岭，面临波光粼粼的柏坡湖，湖光山色，秀美绮丽。馆区内主要景点有：中共中央旧址、西柏坡纪念馆、西柏坡纪念碑、石刻园等。

纪念馆于 1978 年建成，2003 年进行了扩建，总建筑面积 6 125m²。纪念馆依山就势建设，原为两进院落式布局，扩建后将第二进院落改建为半景画厅，并在其后扩建了第三进院落。扩建仍保持了与山体环境紧密结合的院落式布局，突出运用了自然环境的采光通风节能降耗效果。其特点是：①密切配合展陈要求进行改建，重点

Xibaipo is a beautiful village of Pingshan County in Hebei Province, where the Central Committee of Communist Party of China (CPC) lived and directed famous Three Main Campaigns including the Liaoxi Shenyang Campaign, the Huaihai Campaign and the Beiping Tianjin Campaign during the period from May 1948 to March 1949. It is called "The Last Village Headquarter of the Liberation War (Chinese Civil War)". The Second Plenary Session of the 7th Central Committee of CPC was held here in March 1949 and soon after the Session the Central Committee of CPC moved to Beijing from Xibaipo and founded New China, so Xibaipo is regarded as the footstone of the Republic–New China Stemming from here.

The Xibaipo Memorial area is located in the east of the Xibaipo village, backing on Baipo Ridge and facing the rippling Baipo Lake, with beautiful scenery. The main sites in the memorial area include the Site of the Central Committee of CPC, Xibaipo Memorial, Xibaipo Monument, and the Stone Engraving Park.

The memorial was completed in 1978 and renovated in 2003, with its floor area of 6,125 square meters. The memorial was built to the embrace of the mountain. Its original layout was two-courtyard layout, the second courtyard was transformed to half-panorama hall after renovation and the third courtyard was built additionally. The courtyard layout closely combined with the mountain environment was kept and the effect of lighting, ventilation, energy conservation and consumption reduction of natural environment

01

01. 纪念馆旧影（摄于 1978 年）

对页
02. 总平面
03. 平面
04. 纪念馆外景（摄于 2004 年纪念馆改扩建后）

01.Outdoor scene of Xibaipo Memorial after its completion in 1978

Opposite page
02. Master plan
03. Plan
04. Outdoor scene of Xibaipo Memorial after its renovation in 2004

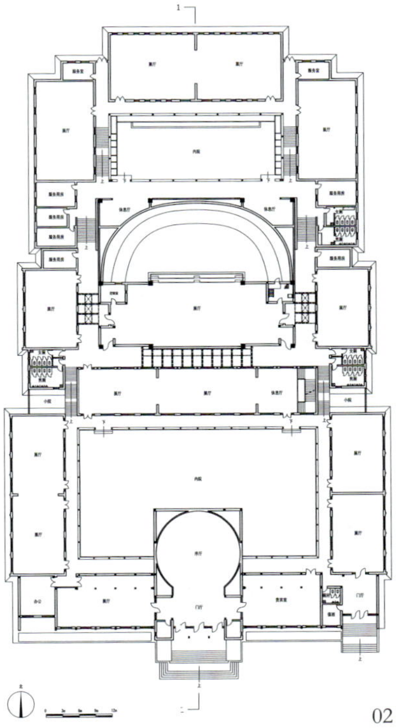

02

总平面图 Master plan 03

04

解决了序厅、半景画厅的大空间展陈需要。展厅及过渡空间也以周到、细致的处理，达到了与展陈的完美结合，确保了展出的最佳效果。②进一步理顺了展出流线，结合庭院将原来分散的展室，整合为 10 个较大的展厅，使主展线更为清晰、顺畅。③保持了纪念馆原有风貌。原建筑在几十年的展出过程中，其朴实无华的形象已深入人心，保持了原有建筑风貌和屋顶群落组合关系，使建筑文化得以延续。④改善了展陈的环境条件。通过改进围护结构性能、增设消防设施及空调系统，大大提高了展出环境条件和参观展览的舒适度。

was obviously utilized during the renovation. The features of the new memorial are: ① renovation closely conforms to the exhibition requirements and primarily meets the large-space exhibition requirement of preface hall and the half-panorama hall. The exhibition hall and the transition space are considerably and carefully treated to perfectly be combined with the exhibitions, thus to ensure the optimum exhibition effect. ② Further rationalize the exhibition line through combining the original scattered exhibition rooms into ten large exhibition halls , making the main exhibition line clear and smooth. ③ Maintain the original style of the memorial. The modest image of original building has enjoyed popular support during its exhibition for scores of years. The maintenance of original building style and the combination relation of roofs extend the building culture. ④ Improve the exhibition environment conditions. The exhibition environment conditions and the comfort for visiting the exhibitions are greatly improved through improving the performance of enclosure structure, and adding fire-fighting facilities and air conditioning system.

05

06

07

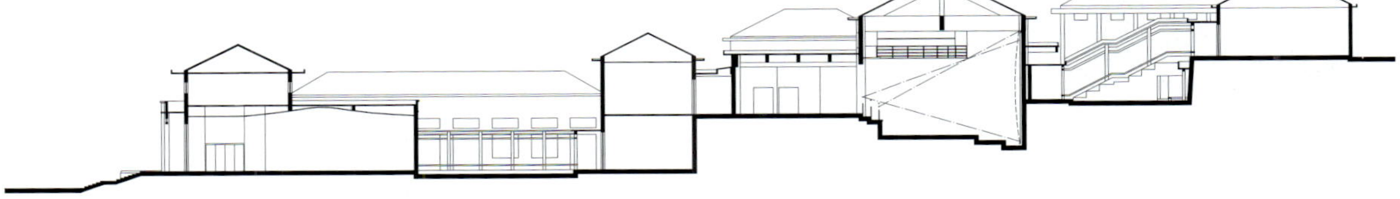

08

07. "三大战役" 展厅
08. 剖面

07. Exhibition hall of Three Main Campaigns
08. Section

对页
05. "两个务必" 庭院
06. 中共中央旧址——毛泽东办公室及寝室

Opposite page
05. Courtyard of the "Two Musts"
06. Site of the Central Committee of CPC—— Mao Zedong's office and dormitory

09

11

10

12

09. 序厅　　09.Preface hall
10. “走进西柏坡”展厅　　10.Exhibition hall of Entering Xibaipo
12. “三大战役”半景画厅　　12. Half-Panorama hall of Three Main Campaigns

13

14

11. “土地会议”展厅
13. “筹备建国”展厅
14. “继往开来”展厅

11. Exhibition hall of Land Reform Conference
13. Exhibition hall of Preparation for Founding of New China
14.Exhibition hall of Carry Forward Our Cause into the Future

Shanxi Museum

山西博物院

山西博物院位于太原市汾河西岸，始建于1919年，初名山西教育图书博物馆，后屡次更名，1953年起称山西省博物馆。2004年新馆建成，定名为山西博物院，荣获“2006年度中国建筑工程鲁班奖”。

山西博物院新馆于2001年8月奠基，2004年12月竣工，2005年9月26日正式对外开放。项目占地面积112 000 m^2，建筑面积51 000 m^2，展厅面积10 000 m^2。总平面采用了中国传统的轴线对称的构图手法，东西轴线为主轴线，南北轴线为次轴线。4层高的主馆位于主次轴线的交会处，形成庄重对称的格局。4座角楼衬托着雄伟的主馆。

主馆4层方正规矩，逐层向外斜挑，雄浑深远，以现代科技手段表现并深化了古建筑的神韵，体现了古人“如鸟斯革，如羽翚斯飞”的审美取向。主馆的主题形象，被赋予了“斗”和“鼎”的寓意。“斗”象征丰收喜悦，“鼎”象征安定吉祥，表达了“和平发展”的象征意义和美好祝愿。外部空间设计采用传统的院落组合手法，生成“太极中央，四面八方”之势，融天、地、人、时、

Shanxi Museum is located on the west bank of Fenhe River. Growing out of Shanxi Educational Book Museum founded in 1919, the museum changed its name many times and was named Shanxi Province Museum in 1953. In 2004, it was finally named Shanxi Museum upon the completion of the new museum. It has won Lu Ban Award for Construction Projects in China in 2006.

The construction of the new Shanxi Museum was laid a foundation in August 2001 and completed in December 2004. On September 26, 2005, the new museum was officially opened to the public. It covers an area of 112,000 square meters and a floor area of 51,000 square meters. Its exhibition halls cover a total area of 10,000 square meters. Its general layout adopts the traditional Chinese axisymmetrical graphics, with the east-west axis being the principal axis and the north-south axis being the secondary axis. The four-storey main building is located at the junction of the two axes with four corner towers as its important setoffs, forming a symmetrical and solemn layout.

From the bottom storey upward, each upper storey of the main building has four exterior walls that slopes outwards much further than the lower storey, which not only bears the style of ancient Chinese buildings, but also fully displays a perfect application of modern technological means to display and even shed new light on the charm of ancient architecture. The theme of the main building has integrated the meaning of "Dou and Ding". The former, a kind of ancient grain measuring tool, represents harvesting happiness, while the latter, a kind of ancient cooking vessel, represents stability and auspiciousness. The design expresses the symbolic meaning of peaceful development and best wishes. The exterior of the building is designed by means of the

01

02

本页
01. 博物馆全景
02. 博物馆旧影（摄于1953年）

对页
03. 历史文化浮雕列柱

01. Display
02. Historical photo (1953)

Opposite page
03. Sculpture column of Historical Culture

04

空于一体，利用水池、叠水、雕塑、室外展品等景观设计创造文化氛围，从而达到空间的层层递进、逐步加深。内部空间先抑后扬，节奏鲜明。中央大厅仿自应县木塔，宽敞明亮，气势恢弘，建筑的意义在这里得到升华。

combining techniques of traditional Chinese compounds. The use of ponds, cascades, sculptures and outdoor exhibits has helped create a cultural atmosphere and add a sense of hierarchy, which may result in the visitors' entrance into a tangible history. Likewise, the interior of the building is also designed with profound historical and cultural connotations by means of sculptures, murals and different patterns.

04. 展陈
05. 土木华章 · 悬空寺模型局部
06. 展厅大厅藻井

p98-101 页
07. 展厅大厅
08. 展陈室

04. Exhibition
05. Construction Achievement – part of the Model Suspended Temple
06. Caisson in exhibition hall

p98-101
07. Exhibition hall
08. Exhibition room

05

06

咨询服务

Taihang Eighth Route Army Memorial

八路军太行纪念馆

01

八路军太行纪念馆位于山西省武乡县城西凤凰山麓，是全国唯一全面反映八路军八年抗战历史的大型专题纪念馆，也是集宣教、收藏、科研、旅游诸功能于一体的综合性红色旅游经典景区。

馆区占地面积 148 000 m²，建筑面积 19 000 m²，由八路军抗战史陈列馆、窑洞战景观、八路军将领组雕、八路军胜利广场、八路雄风碑林公园、八路军抗战纪念碑、“和平颂”主题公园 7 个部分组成。纪念馆坐北朝南，采用中国传统建筑的“中轴对称”布局形式，主馆（八路军抗战史陈列馆）位于整个建筑群的中轴线北部，平面图为“工”字形布局。东馆和西馆体量一致，对称围合在主馆下方两侧。主馆和东西馆高差 3 m，进入馆内全部拾阶而

Taihang Eighth Route Army Memorial lies in the west of Fenghuang Mountain in Wuxiang County of Shanxi Province, and it is the only museum that shows the eight-year history of the Eighth Route Army's Anti-Japanese War in China. As a synthetic red tourism site, it combines education, collection, scientific research as well as tourism as a whole.

The whole site covers an area of 148,000 square meters and the floor area is 19,000 square meters. The site is made up of seven parts, namely the exhibition hall of Anti-Japanese War's history, the cave war show, the sculptures of the Eighth Route Army's generals, the victory square, the stele forest park, the monument of Anti-Japanese War

03

01. 八路军纪念广场
02. 纵剖面示意

对页
03. 纪念馆总平面

01. Memorial square of the Eighth Route Army
02. Longitudinal section

Opposite page
03. Master plan of the memorial hall

02

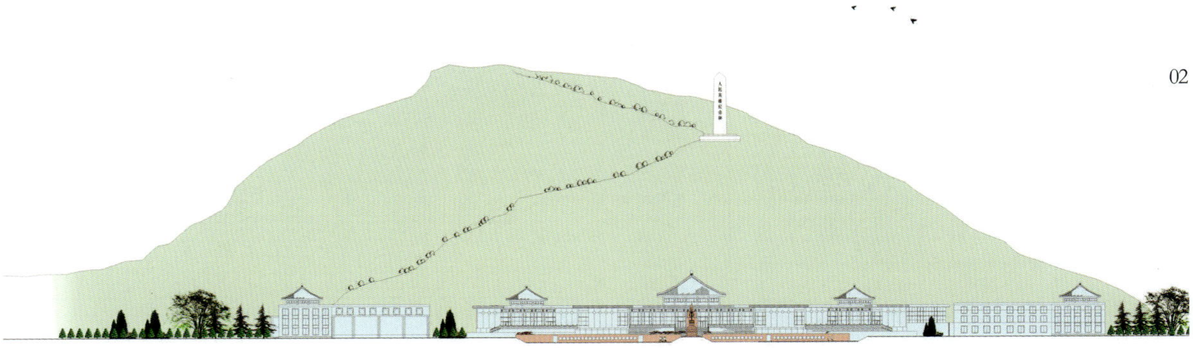

04

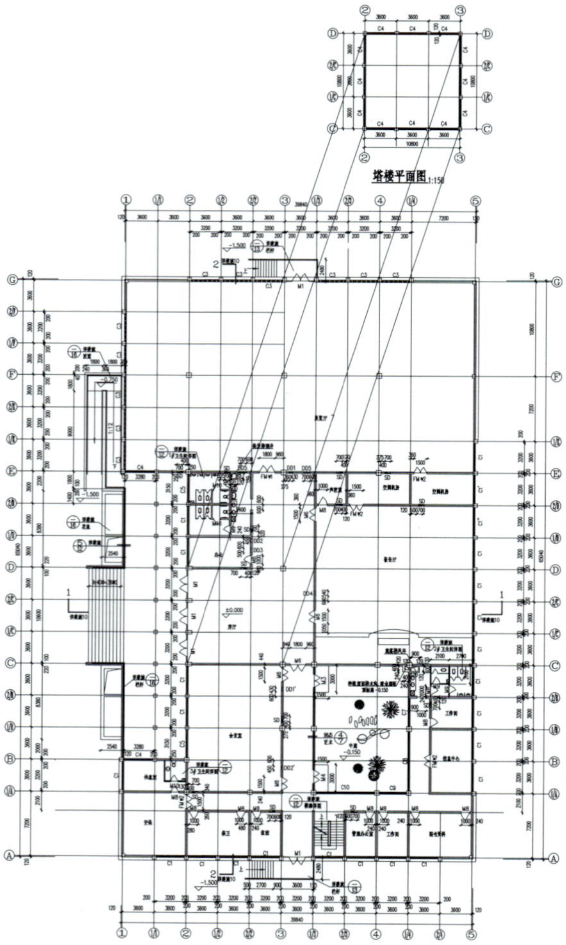

05

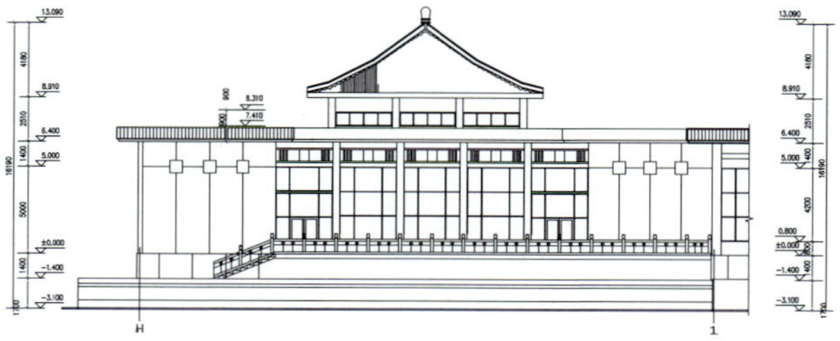

06

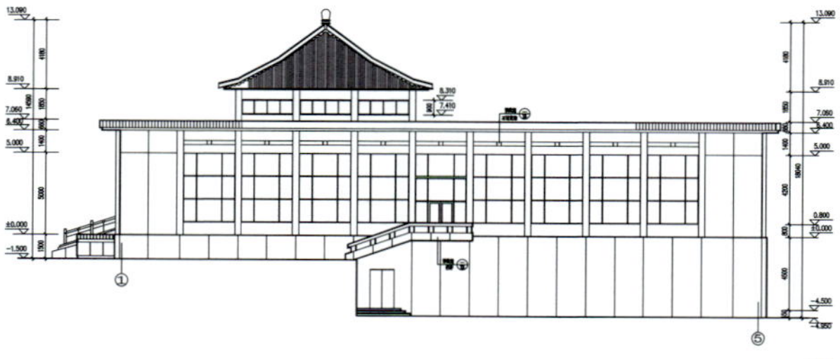

07

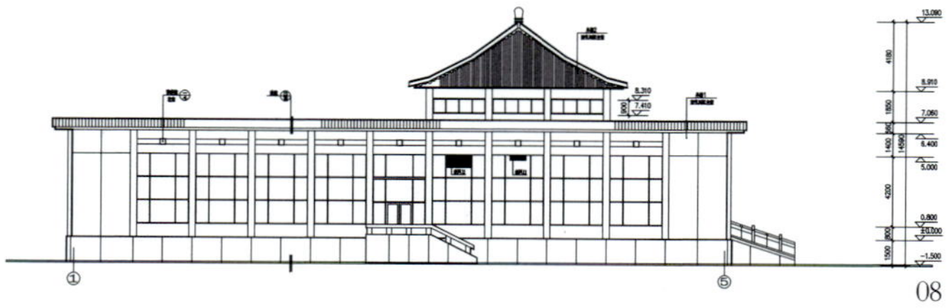

08

本页

04. 纪念馆外景
05. 纪念馆东馆一层平面图
06. 纪念馆主馆立面
07. 纪念馆东馆立面
08. 纪念馆西馆立面

对页

09. 纪念馆外景

04. Exterior
05. Plan of the 1st floor of east hall
06. Facade of main hall
07. Facade of the east hall
08. Facade of the west hall

Opposite page

09. Exterior

09

10

上。三座展馆建筑风格统一，是现代和传统风格有机结合的建筑物。屋顶主要突出传统手法，平面屋顶上分别凸显 5 个四坡型琉璃瓦传统式宝字盖叠加屋顶。屋檐呼应屋顶，1.5 m 女儿墙外立面全部用琉璃瓦装饰围合。墙体风格以现代为主，全部使用暖色调的锈石干挂而成。整个建筑造型古朴典雅、结构精巧，融传统风格和现代气息于一体，气势雄浑，蔚为壮观。

三座建筑中间有花岗岩铺装逾 30 000 m^2 的集散休闲场地，在此中轴线南侧矗立着八路军将领组雕——太行山。组雕主要反映抗战时期朱德、彭德怀等 11 位八路军主要将领的光辉形象，是目前国内展示人物最多、体量最大的领袖群雕。组雕的正南面是纪念馆胜利广场，广场中心八字形纪念碑高耸入云，音乐喷泉水景和绿树草坪交相辉映，与陈列馆、东西展馆、八路军纪念牌形成一个和谐、一致、完整的建筑群。

整个建筑群在苍松翠柏的掩映下，形成了以抗战文化为内涵，园林景色为载体，布局合理、庄重典雅的旅游景区。

11

对页
10. 太行精神光耀千秋——序厅
11. 雕塑——太行山

Opposite page
10. Prelude Hall – The Spirit of Taihang Mountain Will Be Brilliant Forever.
11. Sculpture – Taihang Mountain

and the "peace carol" theme park. The museum faces the south and is constructed according to the Chinese traditional structure's symmetrical principle. The main hall lies in the north of the central axis with the east hall and the west hall lying on its two sides separately. The layout of the structures is just like the Chinese character"工". There is a 3 meters drop in elevation between the main hall and the east and west halls, and visitors can go upstairs to the main hall step by step. The three structures are of the same style that reflects tradition on roofs and modern on walls. The five four-slopes roof covered with glazed tiles jut out over the flat-roof with eaves echoing each other. The walls are all covered with the warm color stone material. The whole structure is simple and elegant in shape, fine and dedicate in structure with imposing manner, presenting a magnificent sight.

Between the three structures, there is a casual area of more than 30,000 square meters covered with granite. To the south of the central axis stands the sculpture of the generals named Taihang Mountain, it mainly reflects the great images of 11 generals during the Anti-Japanese War such as Zhu De, Peng Dehuai and so on. The sculpture is the biggest in size and the most in number of the leader sculpture in China at present. To the south of the sculpture is the victory square, and in its middle stands the monument with the shape of Chinese character"八". Music springs, green trees and lawn enhance their beauty, and all of them compose a harmonious, unanimous and completed structures.

The pines and cypresses set the whole structure off, and all of them form a tourist attraction with Anti-Japanese War as its content, garden view as its carrier and also with rational arrangement and solemn and elegant landscape.

Inner Mongolia Museum

内蒙古博物院

内蒙古博物院为 2007 年内蒙古自治区成立 60 周年的献礼项目之一。工程于 2004 年开始筹备、选址，2005 年 4 月 18 日奠基开工建设，2007 年 7 月 31 日竣工并正式投入使用。

位于呼和浩特市东二环与新华东街交会处的内蒙古博物院，总建筑面积 64200 m^2，总投资约为 7.12 亿元。地上 6 层，高 50 m。首层为文物库房、设备用房区，2 ~ 4 层为陈列展览、观众服务、办公科研区，5 层为办公区，6 层为多功能厅。建筑整体大气独特，设计现代新颖，地域特色鲜明，与周边环境相互映衬，使用功能科学合理。其建筑造型寓意为：美丽辽阔的绿色草原，矗立着刻满历史年轮的毡包。蜿蜒的小路，象征着草原从幽深的远古走来，又向美好的未来走去。舒展的飘蓬，代表着草原上圣洁的哈达，迎接着八方宾客。

Inner Mongolia Museum was built as the salute project to celebrate the 60th anniversary of the foundation of Inner Mongolia Autonomous Region in 2007. The preparation and site selection of the project began in 2004. The construction was started on 18 April, 2005 and completed on 31 July, 2007.

Inner Mongolia Museum is located at the junction of Xinhua East Street and East Second Ring Road of Hohhot, covering an area of 64,200 square meters, with a total investment of 712 million RMB. There are six floors with the height of 50 meters. The storeroom of cultural relics and the equipment area are on the first floor. From the second to the fourth floors, there are exhibition area, visitors service area and scientific research area. The office area and the multi-function hall are on the fifth and sixth floor respectively. The architecture is spectacular, modern and unique in the appearance, scientific and reasonable in the function, with vivid regional features. It represents the scenery of beautiful green grassland and the pure yurts on the grassland. The wandering road implies the long standing history and culture of the grassland. The spacious tent symbolizes the holy Hada greeting the guests from far and near.

01

01. 多媒体触摸屏

对页
02. 博物馆外景
03. 大堂

01. Multi-media touch-screen

Opposite page
02. Exterior of the museum
03. Atrium

02

到目前为止，内蒙古博物院已取得了可喜的成绩，自2007年7月31日投入使用以来，已接待中外参观者300多万人次，旅游旺季日接待量最高可达1.3万人，且已成为社会大众鉴赏珍品、展示成就、提升素养、休闲度假、沟通交流的绝佳场所。

Up to now, Inner Mongolia Museum has achieved delectable results. It has received more than 3 million visitors at home and abroad since it was put into operation on 31 July, 2007. The maximum number of the tourists can be up to 13,000 per day during the peak tourist season. It has also become a great place for the public to feast their eyes on the treasures, cultivate their temperament, relax and communicate with each other.

03

04

05

04. 蒙汉茶吧
05. 展厅

对页
06. 临展——彩棺
07. 第三展厅
08. 皇家风范展览序厅

04. Tea bar of Mongolian and Han nationality
05. Exhibition hall

Opposite page
06. Colored coffin, temporary exhibition
07. 3rd exhibition hall
08. Prelude hall of the exhibition of emperor's style

06

07

08

The 9.18 Historical Museum

“九一八”历史博物馆

“九一八”历史博物馆占地面积 31 000 m²，建筑面积 12 600 m²，主要建筑包括警世钟亭、残历碑、主展馆、胜利纪念碑、办公楼等。

残历碑是博物馆的标志性建筑，也是博物馆最早的建筑，因其造型酷似一本巨大而又残缺的台历而得名。碑体正面右侧是 1931 年 9 月 18 日的日历，左侧则是记述九一八事变真相的碑文。碑面上还有许多大小不等的弹坑，构成呻吟、呐喊的骷髅群，向人们暗示着战争的爆发和灾难的降临。残历碑西南的警世钟亭内悬挂着一口重达 4.2 t 的铜质警世钟，钟的正面铸有“勿忘国耻”四个大字。绕过残历碑是一个大型活动广场，广场西侧刻有捐赠名录和江泽民同志题写馆名的引墙将残历碑与主展馆有机地结合起来。主展馆建筑方案设计与展览方案设计同步进行，打破以往先有建筑、后有展览的传统建馆模式，在建筑形式和功能的设计上充分体现了陈列展览的需求，使建筑和陈列两者完美统一。主展馆出口北侧白色大理石墙面上一组以“奋起”为主题的浮雕，

The 9.18 Historical Museum covers an area of 31,000 square meters and the floor area of the museum is 12,600 square meters. The main building includes Warning-Bell Pavilion, Broken Calendar Monument, the main exhibition hall, Victory Monument and an office building, etc.

Broken Calendar Monument is the landmark building of the museum, and also the oldest building in the museum. It is named Broken Calendar Monument because its shape resembles a large but broken calendar. On the right side of the monument is a calendar of September 18, 1931. The left side is the inscription of September 18th incident. There are many different sizes of craters on the surface of the monument. They compose a group of skulls which are groaning and crying. It implies the outbreak of the war and the catastrophe. Warning Bell Pavilion is on the southwest of Broken Calendar Monument. Warning Bell made of 4.2 tons of copper is hanging in the pavilion. “Don’t forget national humiliation” is carved on the front surface of the bell. Bypass the monument is a large activity square. A wall carved donors’ names and the museum name wrote by Jiang Zemin is on the west side of the square. It integrates the Broken Calendar Monument and the main exhibition hall perfectly. The construction plan and the exhibition plan of the main exhibition building were conducted simultaneously. Therefore, it breaks the traditional exhibition hall building mode that exhibition design follows up construction design. Both architectural style and function design reflect the needs of exhibition display completely, so that the building and the exhibition can be combined perfectly. There is a white marble wall with a group of relief in the theme of “rouse” in the north

01

01. 警世钟亭

对页
02. 东侧雕塑
03. 博物馆入口

01. Warning Bell Pavillion

Opposite page
02. Sculpture on the east side
03. Entrance of the museum

02

以写意手法再现了伟大的抗战历程，既为北侧高 28 m 的抗战胜利纪念碑烘托气氛，又与主展馆入口的“国难”雕塑遥相呼应，彰显了博物馆陈列展览的内涵和主题，同时也使博物馆各个风格独特的建筑成为一个内涵丰富、协调统一的整体。

of the main exhibition hall exit. It reappears the history of the great Anti-Japanese War by impressionistic approach. It not only emphasizes the atmosphere for the 28-meter-high Victory Monument on the north, but also responds the "national crisis" relief at the main exhibition hall entrance. And it highlights the content and themes of display and exhibition in the museum and makes every unique architecture in the museum become a content-rich, coherent whole.

03

04

05

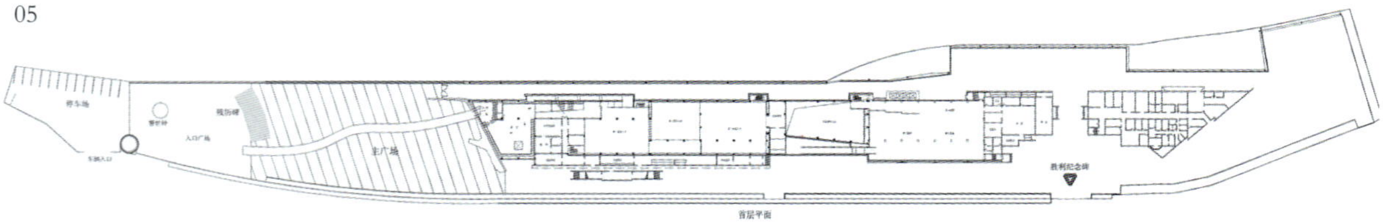

06

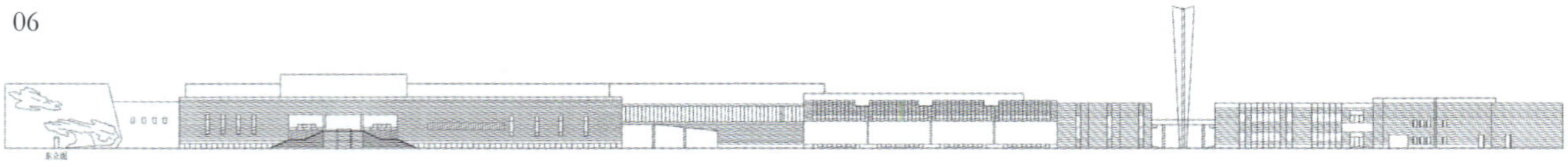

07

04. 东侧外墙面
05. 首层平面
06. 东立面
07. 西立面

对页
08. 博物馆西侧外景
09. 胜利纪念碑
10. 九一八残历碑

04. East facade
05. Plan of the grand floor
06. East elevation
07. West elevation

Opposite page
08. West side exterior of the museum
09. Victory Monument
10. Broken Calendar Monument

08

09

10

Lüshun Museum

旅顺博物馆

旅顺博物馆为大连市属的历史艺术类博物馆，坐落于大连市旅顺口区列宁街 42 号。其前身为日本殖民统治大连时期、始建于 1915 年的满蒙物产陈列所。1916 年改称“关东都督府满蒙物产馆”，1917 年 4 月正式对外开放。1919 年 4 月更名为“关东厅博物馆”。1934 年 12 月定名为“旅顺博物馆”。1945 年 8 月，日本投降，旅顺博物馆由苏联接管，馆名改为“旅顺东方文化博物馆”。1951 年 2 月，苏方将博物馆移交中国政府。1954 年 4 月定名为旅顺博物馆，沿用至今。旅顺博物馆馆舍建筑保存完好，2006 年被评为国家重点文物保护单位。

旅顺博物馆的建筑由主馆和分馆组成。主馆建筑建于 1917 年，是日本殖民当局在沙俄占领时期拟建的“陆军将校集会所”的基础上建筑而成，建筑面积 5 800m²，1918 年竣工。该建筑由日本著名建筑师前田松韵设计，气势恢弘，庄重典雅。楼房主体为砖石木框架结构，主体两层，局部三层，部分有地下室。正立面为左右对称形式，顶部有一穹窿顶塔楼；正面入口

Lüshun Museum is a historical art museum which is located at No.42 Liening Street, Lushun District, Dalian. This museum grew out of Manchu-Mongolia Produce Display Institute built in 1915, when Lüshun was colonized by Japan. In 1916, it was renamed as Guandong (Northeast China) Governmental Manchu-Mongolia Produce Display Institute and opened to public in April, 1917. It was renamed as Guandong Office Museum in April, 1919. In December 1934, it was renamed as Lushun Museum. In August 1945, Japan declared surrender and Lushun Museum was took over by Soviet Union. In February 1951, Lushun Museum was returned to China. Today the buildings of Lushun Museum are still well preserved. It has been honored as important cultural relic site under state-level protection in 2006.

Lüshun Museum building is consist of main museum and branch museum. The main museum was built in 1917. On the base of the Russian Military Officers Club planed by Russian, the building was built by the colonial authorities and finished in 1918. The floor area is 5,800 square meters. The structure of the building is formed by brick and wood, including two stories in main body, three stories and some basements in some pars. The facade is symmetric with an arched tower on the top. There is a colonnade, mixed column caps in the front entrance. There are carvings like pilasters, curly grass lines, vertical grass lines, cherry blosson and animals on the wall. The whole building is typical of European modern eclectic style with ancient Greece, Rome and eastern architectural symbol. In the building, it looks very spacious and bright, with arched doors connected. The showrooms have

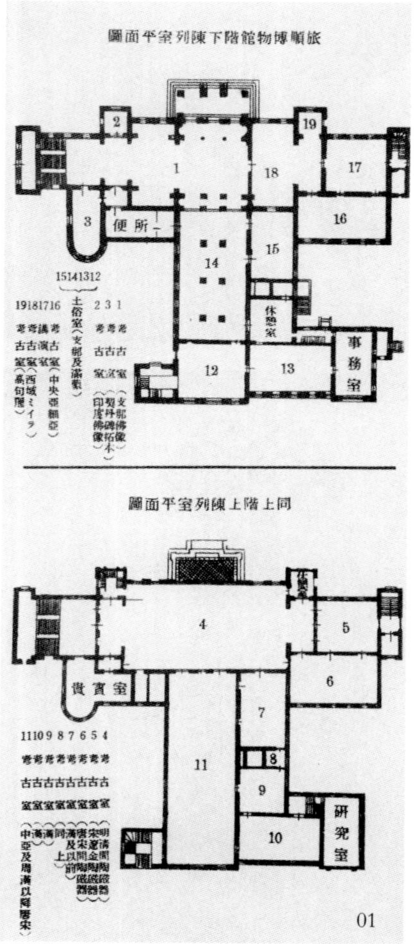

01

02

有柱廊，混合柱头；外墙有贴壁柱，楼面有卷草纹、垂草纹、樱花、兽头等凸起雕饰。整个建筑融古希腊、罗马和东方建筑符号为一体，是典型的欧洲近代折中主义建筑风格。楼宇内部宽敞豁亮，有拱形门相连。玄关为日本德山产花岗石铺就；右方楼阶为美浓产大理石及水户产寒水石；楼梯扶手为意大利墨绿理石贴面。各陈列室设计风格各异，其中二层两个大厅最为华美。北大厅天棚顶为木作式纵横方形分隔，四出斗拱，东西两壁与三楼房间通透相连，造型别致。南大厅天花为拱形顶，南北两壁上方饰有孔雀浮雕图案，既有东方建筑的神韵，又具西方建筑的风采。

旅顺博物馆分馆位于主馆西侧，由原地方建筑框架楼改建而成。建筑工程自 2000 年 1 月开始，2001 年 5 月建成开馆。馆舍建筑的立面设计采用主馆的建筑风格，内部展区则采用现代化展览分区形式，即以大堂为中轴，展厅向两侧铺开，两层 7 个展厅，陈列面积 2 560m^2。

different styles, including the most gorgeous halls on the second floor: North hall's roof is wood-type with vertical and horizontal square separation, four brackets. South hall's ceiling is arched roof, above the north and south nalls decorated peacock embossment. The building is not only a charm of the east building, but also a style of western architecture.

The branch of Lüshun Museum is located on the west of the main hall, and it is built on the original framework of the local building. The construction began in January 2000, and it was completed and opened in May 2001. The premise was designed with the architectural style of the main hall, the internal exhibition area is divided like modern forms. The lobby is the central axis, spreading to both sides, forming seven exhibition halls on two storeys. The whole exhibition area is 2,560 square meters.

03

02. 旅顺博物馆主馆与分馆
03. 博物馆旧影（摄于 1930 年）

对页
01.1937 年旅顺博物馆展览平面图

02. The Main Museum and the Branch Museum
03. Old photograph of the museum (1930)

Opposite page
01. Plan of the Lüshu Museum in 1937

04. 主馆
05. 主馆侧面图
06. 主馆楼面垂草纹、卷草纹雕饰
07. 主馆楼面兽头雕饰

04. The Main Museum
05. Side view of the Main Museum
06. Carving of curly grass lines and vertical grain grass on the facade of the Main Museum
07. Carving of animals on the facade of the Main Museum

08. 分馆
09. 乐池上的孔雀浮雕
10. 主馆二层北大厅木作式天棚顶
11. 二层大厅乐池

08. The Brance Museum
09. Junction with Peacock embossed
10. North Hall’ s Roof: Wooden-type with vertical and horizontal Square Separation
11. Junction in the hall on the second floor

12. 主馆一层与二层全景

对页
13. 馆藏 20 世纪著名画家作品展
14. 主馆前厅右方理石石阶
15. 主馆前厅

12. Panoroma of the first and second floor of the Main Museum

Opposite page
13. Work Exhibition of 20th Century Famous Painters
14. Marble Stairs on the right side of the Main Museum' s atrium
15. Atrium of the Main Museum

13

14

15

Jilin Province Museum

吉林省博物院

吉林省博物院是我国有代表性的重要博物馆之一，是文物收藏保护、科学研究、陈列展示的重镇。1951 年，该馆筹建于松花江畔吉林市，1952 年正式开放，1954 年随省政府迁至美丽的城市长春。

半个多世纪以来，伴随着共和国前进的脚步，吉林省博物院也在不断地发展壮大。现院藏文物十余万件，内容涵盖历代名家法帖、绘画、陶瓷、玉石、青铜以及吉林近现代珍贵文物等 20 多个门类。这里收藏的中国历史书画在国内占据重要席位，其中苏轼《洞庭春色赋》和《中山松醪赋》、张瑀《文姬归汉图》、虞沅《九歌图》、丁观鹏《法界源流图》、董其昌《昼锦堂图卷》等蜚声海内外。夫余、高句丽、渤海文物收藏在东北亚独具特色。传世文物中顶级珍品，世界仅存两件的宋官窑贯耳瓶和宋“松风清节”琴也保存在此。

2007 年，新馆面向全世界公开招投标。2008 年新博物院在长春市净月潭旅游开发

Jilin Province Museum is one of representative museums in China. As a strategic organization for national treasures, it plays an important role in collection and protection of cultural relics, scientific research, and exhibition. With a careful preparation of construction in Jilin, Songhua River bank, it was officially opened to public in 1952, and in 1954 it was moved to the beautiful city of Changchun along with the provincial government.

For more than half a century, Jilin Province Museum is growing stronger and stronger along with the increasing power of PRC. At present, there are more than 100 thousand cultural relics stored here including models of calligraphy by famous experts in various dynasties, drawing, china, jade, bronze, and cultural treasures in modern age, which can be classified into 20 categories. It is a leading museum in collecting the Chinese historical calligraphies and paintings. There are several world famous works such as *Appreciation to Dongting Lake's Sping Scene* and *Appreciation to Zhongshan Wine* by Su Shi, *Wen-chi Return to Han* by Zhang Yu, *Nine Songs Chart* by Yu Yuan, *Dharma Realm Origin Chart* by Ding Guanpeng, *Picture Scroll of Zhoujin Hall* by Dong Qichang. The cultural relic collection of Fuyu, Koguryo and Bohai Sea is unique in Northeast Asia. Among the millenniums-long top cultural relics, the official ware of vase with tubular handles and the instrument of "Songfeng Qingjie" made in Song dynasty are both preserved in the museum.

01

01. 院藏青铜镜展厅契丹纹八角铜镜

对页
02. 老院外景
03-04. 内景

01. Octangle bronze mirror with Qidan grains displayed in the Bronze Mirror Gallery

Opposite page
02. Exterior of the old hall
03-04. Gallery

02

03

04

05

05. 新院鸟瞰图
06. 新院手绘效果图

05. Aerial view of the new hall
06. Hand painted rendering of the newhall

对页
07. 新院夜景
08. 张伯驹、潘素夫妇捐献书画艺术展合图

Opposite page
07. Nightscape of the new hall
08. Picture of Zhang Boju and Pan Su' s Donating Painting and Calligraphy Exhibition

区奠基并破土动工。新院在功能布局、基础设施设备、陈列展览、文物收藏与保护等各方面都将达到崭新水平。建成后的博物院是一个通透式的文化主题花园，集文化欣赏、休闲娱乐于一体。面积 32 000 m^2 的新院与市中心 16 000 m^2 的老馆遥相呼应，必将成为吉林省文化大繁荣的一个美丽景观。

In 2007, the new museum invited public bidding in world-wide scale. In 2008, the construction of the new museum was started in Jinyuetan tourism development zone in Changchun. It will reach a new level in layout, infrastructure, equipments, exhibition, display, the collection and protection of cultural relics and so on. After completion, the museum will be a transparent culture-themed garden, integrating culture appreciation and entertainment. The new museum covering an area of 32,000 square meters will coordinate with the old one which is located in the town centerwith an area of 16,000 square meters, becoming a beautiful landscape of cultural prosperity of Jilin

06

07

08

Jilin Natural History Museum

吉林省自然博物馆

吉林省自然博物馆新馆于 2003 年开工建设，2007 年竣工，占地面积 4800 m^2，建筑面积 16300 m^2。其中，展区面积 7000 m^2，标本库房面积 3000 m^2，办公区面积 3000 m^2。

博物馆建筑呈流线型，采用了非对称动态旋转平面布局。中心主展示空间为椭圆形平面，围绕其布置的辅助部分则为渐开的弧形。外部空间延续建筑体型动势，通过平台、铺地、水池、叠山、绿地等处理，形成相互呼应的、向外发散的旋转环境格局。

博物馆建筑主轴方向与主要公路成一定角度，使得建筑的曲面造型由于观看视角的变化而呈现丰富的不定性。同时，博物馆主入口的巨大玻璃体量从建筑的主轴方向偏转出来，与主要公路平行，上面设置长 23 m 的恐龙雕塑，为自然历史博物馆明显标志。

The new Jilin Natural History Museum started to be constructed in 2003 and was completed in 2007. It covers an area of 4,800 square meters, and the floor area is 16,300 square meters, with the exhibition area of 7,000 square meters, specimen depot area of 3,000 square meters and office area of 3,000 square meters.

This architecture is streamline and arranged with asymmetric dynamic rotating plane layout. The central main exhibition space is elliptic planeand the auxiliary parts surrounding it are gradually opened arc. The exterior space extends the pose of building geometry, through flat roof, pavement, pond, rockery greenbelt, etc., forming the responding and forth-emanative rotating environment patterns.

The principal axis direction of museum architecture forms a certain angle with the major road; therefore, the surface modeling of the architecture presents an abundant inconstancy as a result of the varieties in visual angles. Simultaneously, the huge vitreous body at the major entrance of museum deflects outwards from the principal axis direction of architecture, paralleling to the major roads, on which the dinosaur sculpture with length of 23 meters is mounted as the distinct sign of the museum.

01

01. 北侧

对页
02. 南侧
03. 剖面
04. 主次入口立面
05. 北侧

01. North side

Opposite page
02.South side
03. Section
04. Elevation of the entrances
05. North side

02

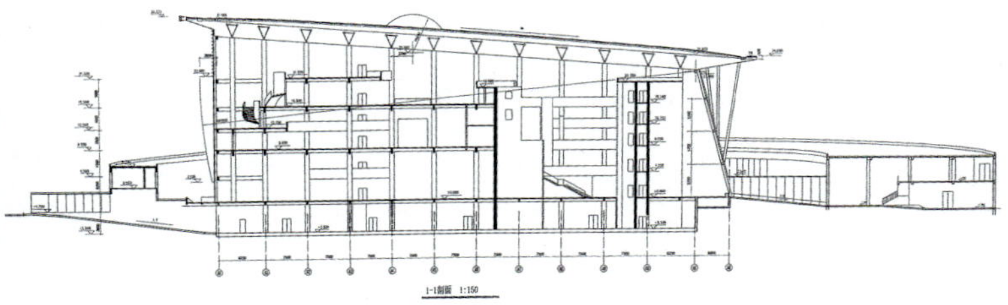
03

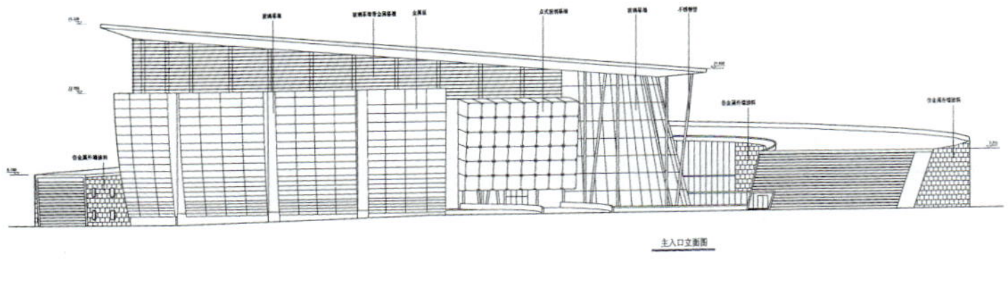
04

05

06

建筑内部空间按照博物馆功能要求分展示区、标本库房区、办公区，并且三者相对独立。展示区共分6层，1、2层陈列古生物化石标本，3层陈列鸟类、兽类标本及景观，4层陈列蝶类标本，5、6层陈列长白山生态景观。

参观人流将首先沿着象征着长白林海的宽大楼梯拾阶而上，途中将领略长白山最具特色的垂直分布带，徜徉于展厅中，观众深深地融入景观中，并不断感受场景和展品的变化。展馆以标本为中心，分类布置，形成人流的主要引导方向。

The interior space of the architecture is divided into exhibition area, specimen depot area and office area according to the functional requirements of museum, relatively independent. There are 6 floors in the exhibition area, in which the first and second ones exhibit the specimen of paleontology fossils, the third one for the specimen and sights of Aves and animals, the 4th one for the specimen of butterflies, the fifth and sixth for the ecological landscape of Changbai Mountain.

The visitors will go up by steps along the wide stairs representing the forests of Changbai Mountain first, appreciate the most characteristic zones at different elevations of Changbai Mountain in the midway, roam in the exhibition hall, deeply integrate into the landscape and continually feel the changes in scenes and exhibits. The exhibition focuses on the specimen, lays by classes and forms the major guiding direction of visitor streams.

06. 北侧

06. North side

The Provincial Museum of Heilongjiang

黑龙江省博物馆

黑龙江省博物馆馆舍建于 1906 年，最初为莫斯科商场，1923 年开始作为研究机构和陈列场所。该建筑高两层，带地下室，平面为梯形，由三个标准段和两个转角联系体组成。楼体采用流畅曲线的墙墩、窗洞和隆起的方扁形穹顶，取得优美的造型和丰富的天际线；米黄色墙面，暗红色穹窿顶，属于典型的巴洛克式建筑。

1986 年该建筑被哈尔滨市政府列入一类保护建筑。这类建筑在哈尔滨近代建筑中已寥寥无几，经过近一个世纪的风雨沧桑，楼体建筑破损较重，作为省级重点文物保护单位，急需保护维修，恢复原始风貌后，将成为哈尔滨冰城一处靓丽的景点。

黑龙江省博物馆始创于 1923 年，初名“东省文物研究会陈列所”，后更名“东省特别区文物研究会博物馆”、“松江省科学博物馆”等，直到 1954 年才正式改称黑龙江省博物馆，1962 年郭沫若先生题写馆名。本馆是新中国成立后首批省级综合性博物馆，承担着黑龙江地区历史文物、自然标本的保护、征集、采集以及研究、陈列等工作。经过几代人的不懈努力和积累，现有馆藏历史文物、艺术品 3 万余件，动植物和地矿标本 8 万余件。本馆具有独特的馆藏优势和陈列风格，长期对外展出的基本陈列有“黑龙江历史文物陈列”、“自然陈列”、“海洋世界”，每年还有不断更新的临时展览推出，为黑龙江省文博事业的发展做出了一定贡献。

The building of the Provincial Museum of Heilongjiang was built in 1906, which was originally a Moscow Shopping center. It started serving as a research institution and exhibiting hall since 1923. The building has a two-floor height, equipped with basement, having a trapezoid-shaped plane which is composed of three standard segments and two corner junctions. The body of the building has the kind of wall pier and window with fluid curves, and raised vaults shaped as fat rectangle, so as to achieve the elegant style and maintain multiple skylines. Besides, as the typical features of a baroque building, the beige wall and the maroon vault can be seen everywhere.

In 1986, the municipal government of Harbin figured this building into first class protected buildings. As of now, this kind of building has become one rarity of the latter-day buildings in Harbin. Through nearly one-century vicissitude, the building body has been badly damaged. As a historical and cultural relic under provincial protection, it is badly in need of restoration to recover its original looks, and finally it can become one bright attraction of Harbin ice city.

The Provincial Museum of Heilongjiang was founded in 1923, with an original name as "The Exhibition Hall of Eastern Province Academy of Historic and Cultural Relics", and it was successively renamed as "The Museum of Eastern Province Special District Academy of Historic and Cultural Relics", "The Scientific Museum of Songjiang Province" and so on. Until 1954, it was officially named as "The Provincial Museum of Heilongjiang", and the name was inscribed by Mr. Moruo Guo in 1962. This museum was in the first batch of provincial level comprehensive museums after the founding of People's Republic of China, and it takes the responsibility of protecting, collecting, researching and showing historical relics and natural samples in Heilongjiang. With the aid of the unremitting efforts and accumulations of generations, the current collection of the museum has over 30 thousand historical relics and artworks, with over 80 thousand creatural, botanical and mineral samples. The museum has an extraordinary resource advantage and exhibition style. The fundamental collection which is long-term exhibited to the public includes "The Exhibition of the Historical Relics in Heilongjiang", "Natural Exhibition" and "Sea World". There are also constantly updated temporary exhibitions displayed annually, which make great contributions to Heilongjiang's cultural and museum undertakings.

01

01. 展品

对页
02. 外景
03. 博物馆主入口
04. 院落

01. Exhibits

Opposite page
02. Exterior
03. Main entrance
04. Courtyard

02

03

04

05

05-08. 展陈厅

05-08. Exhibition hall

06

07

08

Iron Man Wang Jinxi Memorial

铁人王进喜纪念馆

大庆铁人王进喜纪念馆作为我国第一座工人纪念馆，已历经 40 个春秋。如今，它已成为我国最具石油特色的文化景观和弘扬民族精神的神圣殿堂。

大庆油田先后于 1971 年 7 月、1975 年 5 月、1991 年 11 月和 2003 年 2 月兴建、改建、扩建和迁建铁人王进喜纪念馆。新建的铁人王进喜纪念馆位于大庆市让胡路区，占地面积 11.6 万 m²，于 2006 年 9 月 26 日油田发现 47 周年之际正式开馆。馆名由温家宝总理亲笔题写。

纪念馆建筑外形由“工”、“人”二字组合而成，鸟瞰呈“工”字形，侧看为“人”字形，象征这是一座工人纪念馆。其中，主体建筑高 47m，正门台阶 47 级，寓意铁人 47 年短暂而不平凡的人生。建筑顶部为钻头造型，象征着大庆油田锐意进取，奋发向上，开创未来。该建筑曾荣获中国

As the first workers' memorial of our country, Iron Man Wang Jinxi Memorial has been there for 40 years. Now it has become the cultural landscape with the most oil features in our country and the sacred palace which promotes the national spirit.

Iron Man Wang Jinxi Memorial was first built in July 1971, rebuilt in May 1975, again rebuilt and extended on the basis of the original exhibition hall in November 1991, and afterwards moved in February 2003. The newly-built Iron Man Wang Jinxi Memorial is located in Ranghulu District, Daqing City. The memorial covers an area of 116,000 square meters. Premier Wen Jiabao, personally inscribed the name of the Iron Man Wang Jinxi Memorial. On September 26, 2006, the occasion of the 47th anniversary of Daqing oil field's discovery, it was formally opened to the public.

The external design of the memorial contains two Chinese characters "工"and "人". From above, the character "工" can be seen, while from the side, the character "人" is visible. "工人" means "the worker", this symbolizes it is the workers' memorial hall. Among it, the main construction is 47 meters high and has 47 steps to the main entrance, signifying the short but extraordinary 47 years of Wang Jinxi's life. The top of the building is in the

01

01. 纪念馆旧影（摄于 1975 年）

对页
02. 夜景
03. 鸟瞰
04. 总平面

01. Old photograph of the memorial (1975)

Opposite page
02. Nightscape
03. Aerial view
04. Master plan

02

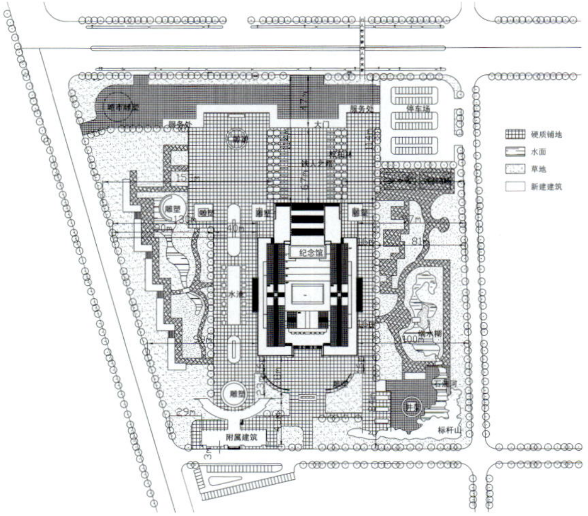

04

经济技术指标

规划总用地：12.4万m^2

总建筑面积：2.27万m^2

其中纪念馆：1.9万m^2

办公楼：0.17万m^2

服务、售票：0.20万m^2

建筑密度：0.72

平均层数：2.38

容积率：0.17

绿地率：59.67%

03

建设工程鲁班奖，被誉为大庆市“最具标志性的建筑”。

纪念馆正前方矗立着高 6.5m、重 60t 的铁人手扶刹把雕像，两侧是主题雕塑“崛起”和“奋进”，很好地诠释了大庆精神、铁人精神的深刻内涵，做到了主体建筑、馆区环境与馆内布展的完美结合，被国内著名陈展专家誉为名副其实的国家一级馆、“人馆合一”的精品工程、纪念馆建设的翘楚之作。

shape of a drilling bit, symbolizing Daqing oil field's great rise, spurting bravely with determination and creating a new future. This construction has won the honorable title of "Luban Award", which is the national construction project's highest award and it is honored as "the most symbolic building" of Daqing.

In front of the memorial hall, the statue of Iron Man with hands clasping the brake crank, which is 6.5 meters high and weighs 60 tons, has been erected on the platform. On the sides are theme statues *Rising Abruptly* and *Spurting Bravely* which make a profound annotation of the rich connotations of Daqing spirit and Iron Man spirit. It has accomplished the perfect unity of the main building, the environment of the memorial area, display arrangement the memorial. It is honored as a veritable national class-A museum, the excellent project of the unity of man and memorial, an outstanding work of the memorial construction by famous domestic exhibition experts.

05

06

07

08

05. 纪念馆外景

对页
06-08. 铁人手扶刹把雕像

05. Exterior of the memorial

Opposite page
06-08. The granite statue of Iron Man with hands clasping the brake crank

09

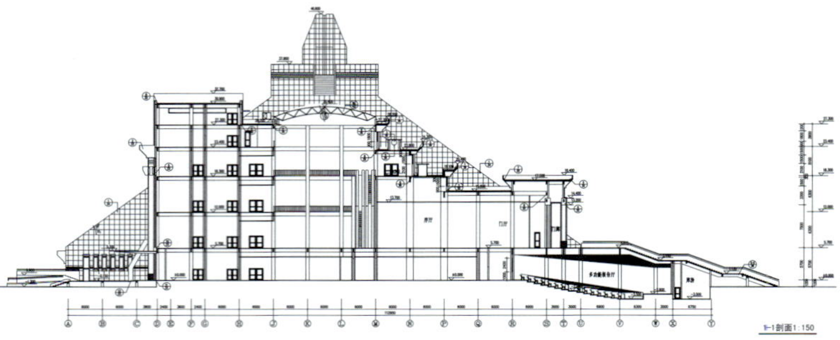

10

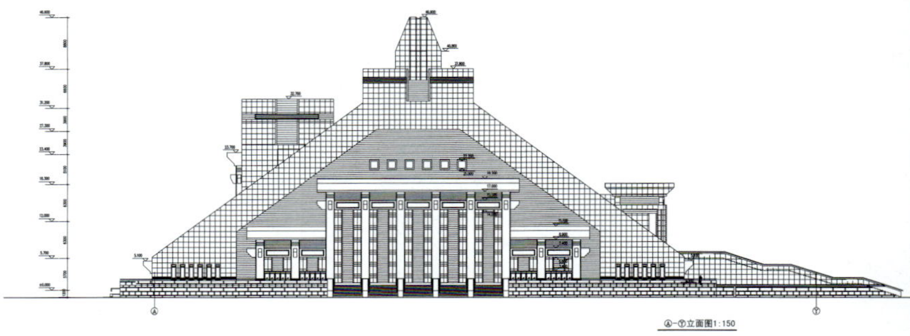

11

12

13

14

13. 第四展厅
14. 阳光大厅丰碑雕塑

对页
09. 第一展厅
10. 剖面
11. 立面
12. 第三展厅

13. The Fourth Exhibition Hall
14. The monument in the Sunshine Hall

Opposite page
09. The First Exhibition Hall
10. Section
11. Elevation
12. The Third Exhibition Hall

The New Shanghai Museum

上海博物馆新馆

上海博物馆新馆于 1993 年 8 月开工，1996 年 10 月 12 日全部建成开放。新馆地处市中心人民广场中轴线南端的黄金地段，与上海市政府人民大厦南北相对。

新馆占地 12 552 m^2，建筑面积 39 200 m^2，地上五层，地下两层，高度 29.5 m。新馆的建筑造型，把中国传统文化的精髓及典雅大方的建筑格调和现代科技创新形式相结合，既有时代气息，又反映了博物馆的艺术特色。新馆建筑是方体基座与圆形出挑相结合，具有中国传统的“天圆地方”之意，整个外形犹如一尊庄重古朴的巨型古代青铜器；流线型的圆顶，向四面开放的拱门，显示了跨世纪之形象；拱门与外墙花岗石贴面的雕刻花纹取材于青铜器纹饰，以现代手法表达出阴阳与凹凸的交相生辉；南大门外的 8 尊汉白玉石雕塑，每件高 2.8 m，重近 20 t，仿制于汉至唐代的文物藏品，融合了建筑功能的文化内涵。

新馆设有中国古代青铜、陶瓷、书法、绘画、雕塑、印章、钱币、少数民族工艺、玉器、家具等 11 个基本陈列馆和 3 个特别展览

The New Shanghai Museum started to build in August ., 1993 and was completed on October 12th, 1996. It is located in a golden place at the south end of the medial axis of the People's Square, facing the People's Building (Shanghai Municipal People's Government).

The site area of the museum is 12,552 square meters, and the floor area is 39,200 square meters. There are five floors aboveground and two floors underground. The height is 29.5 meters. Combining the elegance of the Chinese pattern of the traditional culture with the elegant and graceful architectural style and the innovative form of modern technology, the museum not only shows the characteristic of the times, but also reflects the art character of the museum. The rectangular platform and the circular roof mean the old Chinese ideology of "the square earth is under the round sky". The appearance of the building seems like a huge sacred bronze vessel of primitive simplicity. Streamlined dome, surrounded by open arches shows the image of the next century; arches and carving patterns of the exterior granite veneer are drawn from the bronze decoration, express the Yin and Yang and the cross-brilliance of the concave and convex in a modern style. There are eight white marble sculptures outside the south gate, each of which is 2.8 meters high and weights approximately 20 tons, are the imitations of the relic collections from the Han Dynasty to the Tang Dynasty, and also integrate the cultural connotation into the architectural function.

The New Shanghai Museum has eleven galleries (the galleries of Ancient Bronze, Ancient Chinaware, Calligraphy, Sculpture, Prints, Coins, Nationalities' Art, Jade and Ming and Qing Furniture) and three special temporary exhibition halls. Besides, there are many advanced facilities and complete functions including the high-definition movie & television center, information center, libraries, international academic report halls and conservation technology

01

01. 博物馆旧影（创建于 1952 年的上海博物馆原址位于南京西路 325 号旧跑马总会）

对页
02. 博物馆正门
03. 入口处的石狮

01.Old photograph of the museum (No.325, West Nanjing Road — old Shanghai Race Club, the old site of the Shanghai Museum, founded in 1952)

Opposite page
02.The main entrance of the museum
03.A Stone lion at the entrance

02

03

06

China Maritime Museum

中国航海博物馆

中国航海博物馆位于上海市浦东新区临港新城，于 2010 年 7 月 5 日正式开馆。博物馆为一栋独立的 4 层建筑，高约 24 m，总面积 46 430 m^2。

本建筑主体结构采用现浇钢筋混凝土框架结构体系。楼、层盖均采用现浇钢筋混凝土梁板结构，局部柱、梁采用劲性柱、梁，在核心区域设置斜撑，以增加高耸帆体周边结构的刚度。

其中，帆体为钢网格结构与钢索张拉结构。此结构特别体现了该建筑极具表现力的屋顶造型，它犹如两张仅在一点上相互接触的弯曲的风帆。这两张风帆有 3 个独立的呈三角形的端点。大型的透明的弧形立面建在风帆之间。该结构总高度大约为 58 m（至风帆顶端）。每个三角形风帆的底部两支点间距大约为 70 m。两张风帆的交叉点也就是立面最高点大约离地 40 m。立面各处宽度不等，但最宽不超过 24 m。高 6.7 m 的玻璃斜立面位于建筑的边缘。

China Maritime Museum is located in Lingang New City of Shanghai Pudong New Area and officially opened to the public on 5th July,2010. As an independent four-storey building, its total height is around 24 meters with a total floor area of 46,430 square meters.

The main structure adopts cast-in-situs reinforced concrete frame, with the floor system using cast-in-situs reinforced concrete slab and girder. Rigid columns and beams are partly used. Diagonal braces are set up in the central area. All the measures mentioned above are used to reinforce the stiffness of the structure around the central sail.

The structure system of this sail construction is a combination of steel grid structure and tensioning steel cable structure.The structure is characterized by its expressive roof shape that resembles two curved sails that are touching each other at only one point. The two sails have three (unattached end points that are arranged in a triangular pattern. Large transparent curved facades are built between these sails. The total height of the structure unattached to the top end is approximately 58 meters. The two base points of each triangular sail are approximately 70 meters apart. The intersecting point of the two sails, namelly, the top of the facade is approximately 40meters above the ground. The facade varies in width with a maximum of approx .24 meters. Inclined glass facades with a maximum height of approx.6.7m are located along the sides of the building.

对页
01. 博物馆夜景
02. 总平面
03. 东立面
04. 北立面

Opposite page
01. Nightscape of the museum
02. Master plan
03. East elevation
04. North elevation

01

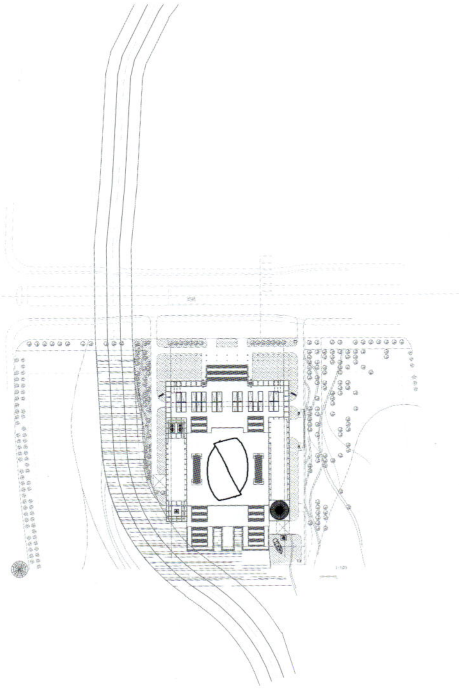

02

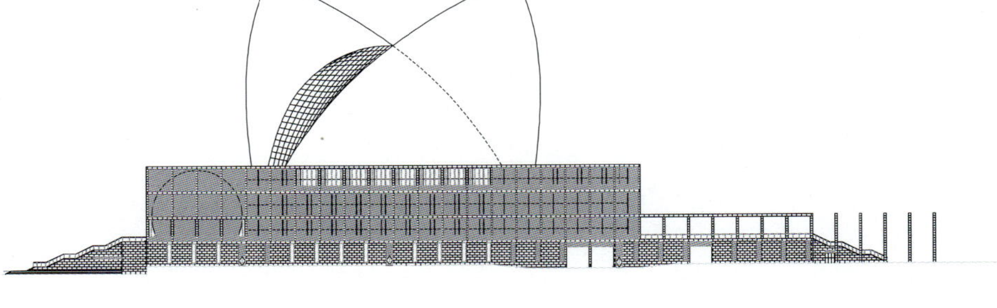

03

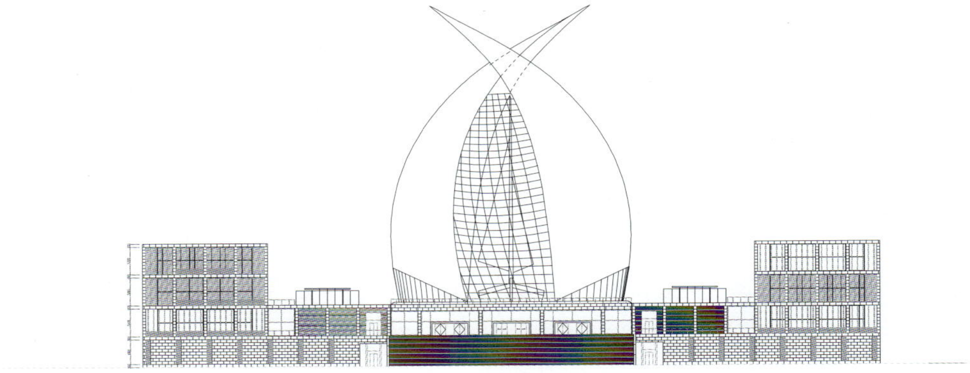

04

05. 博物馆鸟瞰　05.Aerial view of the museum

Shanghai Science & Technology Museum

上海科技馆

上海科技馆的建筑外观美观新颖，充满创新意识和深刻意蕴。建筑整体与上海科技馆“自然·人·科技”的办馆主题充分融合，主体呈现西低东高、螺旋上升的不对称结构。建筑采用弧形平面，螺旋上升的平面将主体覆盖在一片由西向东缓缓升起的巨型翼状屋顶之下，以此极富升腾、前冲姿态的动态造型，隐喻巨龙腾飞、大地崛起。整个建筑分为三个部分：西侧是一个由低到高逐步递增的扇形空间；中部是透明的玻璃卵形大堂（最高处为 42 m，最宽处为 67 m）和中央的黄色球体（直径 18 m），象征着宇宙的无垠、生命的孕育；东侧是 4 层的框架结构，每层 10 m，最高隆起处高出地面 49 m。

整体建筑象征自然、人与科学的演化、发展、提升和腾飞，蕴涵较高的艺术和审美情趣。该建筑因其现代先锋的设计理念、恢弘美观的视觉形态以及突破原水渠保护、卵形铝钛合金单层网壳施工技术、外墙防渗、大面积深基础施工等问题，而获得上海市建设工程“白玉兰”奖、第三届

The building of Shanghai Science & Technology Museum has an attractive and novel appearance and is full of innovative consciousness and profound connotations. As a whole, it fully reflects the theme of "Nature, Mankind, Science & Technology", adopting an asymmetric structure of a climbing spiral low in the west and high in the east. With an arc plane, the climbing spiral covers the main body of the building under a giant-wing-shaped roof that is arising from west to east, so as to create a dynamic, vigorous and rushing pattern, giving the connotation of the oriental dragon soaring up and the land rising. The whole building can be divided into three parts: the western part being a fan-shaped space that ascends and expands gradually; the middle part being a transparent ovum-shaped glass hall (42 meters high at the highest point and 67 meters wide at the widest point) and a central yellow ball (18 meters in diameter) symbolizing the infinite universe and the gestation of life; and the eastern part being a 4-storey framework, with the floor height of 10 meters at each floor and the highest point 49 meters above the ground.

The whole building symbolizes the evolution, development, enhancement, and soaring up of the nature, mankind, and science, with rich artistic and esthetic taste. Because of its modern avant-garde design concept, its majestic visual impact, and its breakthrough in original water channel protection, its ovum-shaped aluminum-titanium alloy single-layer lattice

01

01. 球面特写—— 球面玻璃反射夕阳

对页
02. 科技馆日景
03. 科技馆正面

01. Close-up of the spherical surface-glass reflects the rays of the sunset

Opposite page
02. Museum in the daytime
03. Facade of the museum

02

03

04

詹天佑土木工程大奖和中国建筑工程“鲁班奖”，更成为来往游客争相参观留念的场所，是上海市的又一大标志性建筑。

shell construction technology, leakage-proof exterior wall, and large-area deep-foundation construction, the building has won Shanghai Magnolia Award of Construction Engineering, the 3rd Zhan Tianyou Civil Engineering Award, and China State Construction Engineering Lu Ban Award, and has become a popular tourist site for domestic and overseas visitors and another landmark building in Shanghai.

04. 内景

对页

05. 结构特写——侧面看大玻璃和曲线

06. 结构特写——在大厅看外部曲线

04. Interior

Opposite page

05. Structure close-up — side view of the large glass and curve

06. Structure close-up — view of the external curve

特展厅 四维影院（二层入口） 蜘蛛展 一层展区
Special Exhibition Hall IWERKS 4D Theater (2F Entrance) Spiders 1F Exhibition
机器人世界 信息时代 地球家园
World of Robots Information Era Home on Earth

05

06

Nanjing Museum

南京博物院

南京博物院坐落于风景秀丽的紫金山南麓，东倚中山门，面临清溪河，占地面积逾 70 000 m²。

主体建筑包括历史陈列馆与艺术陈列馆，二者均为仿辽式宫殿建筑，遥相呼应，相得益彰，显示出浓厚的传统文化氛围。历史馆 1936 年开工建设，因历史原因至 20 世纪 50 年代初建成，面积 14 000 m²，其中陈列室 4 800 m²，配套公用建筑 5 300 m²，文物库房 2 900 m²，目前为省级文物保护单位。

艺术馆 1999 年建成，当年即被评为“全国十大精品陈列”，面积近 15 000 m²，一层展厅近 8 000 m²，设珍宝馆、陶艺馆、瓷器馆、青铜器馆、民俗美术馆、国际学术演讲厅、青年雕塑家陈列馆；二层展厅近 7 000 m²，设漆器馆、玉器馆、名人书画馆、织绣馆、现代艺术馆。各馆陈列既自成体系，又相互协调，共同构筑起一座神圣的艺术殿堂。

Nanjing Museum is located at the southern foot of scenic Purple Mountain, leaning on Zhongshan Gate, facing Qingxi River, and it covers an area over 70,000 square meters.

The main buildings are the Historic Exhibition Hall and the Art Gallery. Both of them are the architecture in the style of Liao Dynasty, which manifest the Chinese traditional cultural atmosphere. The construction of the Historic Exhibition Hall started in 1936 and onded in the early 1950s, which is the Provincial Cultural Relics Protection Unit now. The area is nearly 14,000 square meters, including the 4,800 square meters of exhibition places, the 5,300 square meters of supporting public buildings and the 2,900 square meters of cultural relics storeroom.

The Art Gallery was completed in the year of 1999, and in the same year it was awarded as one of the "National Top 10 Elite Exhibits". The area is nearly 15,000 square meters. The Art Gallery consists of two floors. The first floor is nearly 8,000 square meters, including the Ancient Treasure Hall, Earthware Exhibition Hall, Porcelain Hall, Ancient Bronze Hall, Folk Art Hall, International Academic Lecture Hall, and the Youth Sculptors' Works Exhibition Gallery. The second floor is about 7,000 square

01

01. 博物馆旧影（摄于 20 世纪 50 年代）

对页
02. 历史陈列馆
03. 艺术陈列馆

01. Old photograph of the museum (1950's)

Opposite page
02. Historic exhibition hall
03. Art gallery

02

03

2009 年，南京博物院二期改扩建工程开工建设，历史馆与艺术馆即将进行大规模改造。至 2012 年，南博将形成包括历史馆、艺术馆、特展馆、民国馆、数字馆、非遗馆在内的“一院六馆”建置，服务公众能力将会在原有基础上得到更进一步提升。

meters, including Lacquer Ware Hall, Ancient Chinese Jade Hall, the Celebrity Painting and Calligraphy Gallery, the Silk and Embroidery Hall, and the Modern Art Gallery. They make a holy art palace together of Nanjing Museum.

The reconstruction project started in 2009, and the Historic Exhibition Hall and the Art Gralley would be extensively rebuilt. In the year of 2012, Nanjing Museum will have completed the system of “One Museum with Six Exhibition Halls”, that is to say, Nanjing Museum will set up 6 exhibition halls of different themes such as Art Gallery, Chinese Historic Hall, Temporary Display Halls, Intangible Cultural Heritage Halls, Digital Display Halls, and the Halls on History of R.C. Period. Its capacity of serving the public will be improve further than before.

04

05

04. 博物馆二期下沉广场透视图
05. 金镶玉成效果

对页
06. 剖面
07. 博物馆二期改扩建工程鸟瞰

04. Perspective of the sunken square in the reconstruction project, phase Ⅱ
05. Decoration of “Gold and Jade”

Opposite page
06. Section
07. Aerial view of the reconstruction project, phase Ⅱ

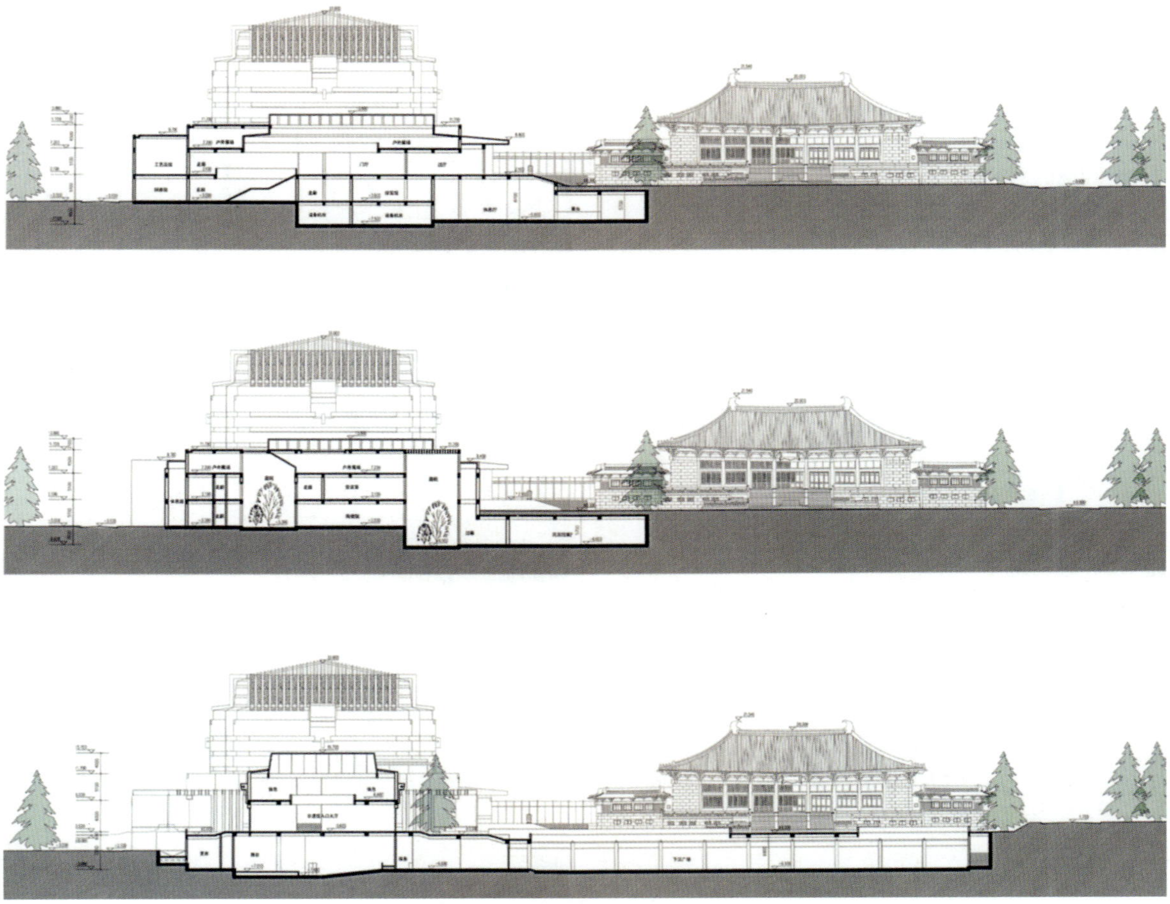

06

07

The Memorial Hall of the Victims in Nanjing Massacre by Japanese Invaders (New Hall)

侵华日军南京大屠杀遇难同胞纪念馆新馆

侵华日军南京大屠杀遇难同胞纪念馆建立在南京大屠杀江东门集体屠杀场地及“万人坑”遗址之上。新馆于 2007 年 12 月 13 日南京大屠杀 30 万同胞遇难 70 周年之际建成开放，占地面积约 74 000 m^2，建筑面积 25 000 m^2。新馆分为展览集会区、遗址悼念区、和平公园区和馆藏交流区 4 个功能性区域，是一座全面展示南京大屠杀特大惨案的纪念性遗址型历史博物馆。

新馆的建筑平面布局为和平之舟，空间布局为化剑为犁。整个建筑设计构思概括为“死亡、和平”。前半部分为“白骨为证、废墟为碑”，用有特色的建筑语言固化历史，封闭的内场环境保护了历史遗址特有的氛围，给观众极强的震撼力；后半部分体现“人类家园、走向和平”，半开放的绿色空间使观众从历史时空回到现实世界，用有深刻寓意的建筑语言来增强人们对战争、和平、人性的想象，突出了对人类和平的向往。

The Memorial Hall of the Victims in Nanjing Massacre by Japanese Invaders was established on the relic of the Jiangdongmen slaughter place and the "Mass Grave". The new hall opened on Dec.13th, 2007, which was also the 70-year-anniversary of the Nanjing Massacre. The site area is 74,000 square meters and the floor area is 25,000 square meters. The building is divided into four functional parts: the exhibition area, the relic memorial area, the peace park and the storage and communication part, which together make the museum a comprehensive memorial relic historic museum that displays the tragedy of the Nanjing Massacre.

The museum's planning means "the boat of peace" and the space means changing the sword into plowshares. The idea can be generalized as "the death and the peace". The first part is "bones as evidence, ruins the monument": The history is represented by the distinctive architectural language while the closed inner courtyard also keeps the historical relic atmosphere and has an earthshaking impact on the visitors. The later part shows the "human's home world and the way to peace", the half-opened green space draws the visitors from history to the reality and promotes their imagination of war, peace and humanity with some meaningful architectural language and emphasizes the expectation for our human's peace.

01

01. 灾难之墙，在黑色的花岗岩石上刻着中、英、日、德、葡萄牙、希腊、荷兰等 12 国文字的“遇难者 300000”。

对页

02. 胜利之墙夜景

03. 纪念馆规划总平面

01. Wall of Disaster, black granite carved with the words “Victims 300,000” in 12 Languages: Chinese, English, Japanese, German, Portuguese, Greek and Dutch

Opposite page

02. Nightscape of Victory Wall

03. Master plan of the memorial

02

纪念馆新馆扩建工程荣获 2009 年度中国建筑工程“鲁班奖”、中国建筑学会建国 60 年建筑创作大奖、教育部优秀建筑设计一等奖、第五届中国建筑学会建筑创作优秀奖等，得到了建筑界专家的认同和好评，认为该建筑“承载悲情，祈求和平，震撼人类心灵”。两院院士吴良镛先生于 2007 年 12 月 27 日来馆参观时评价说：“新馆建筑设计很有气势，也很朴素，用写实的手法表现了中华民族所蒙受的灾难。”

The expansion project of the new hall wins many honors including China State Construction Engineering Lu Ban Award in 2009, the architectural design prize in the 60th anniversary for the foundation of PRC from CIA (Chinese Institution of Architecture), the first-level architectuarl design prize of the Ministry of Education in China and the excellent architectural design prize in the 5th conference of CIA. The new hall also wins acceptance and many reputations from the experts in architecture, who believe the building "holding the pathos, praying for the peace and heartquaking". During his first visit to the new hall on Dec.27th, 2007, Mr. Wu Liangyong "the academician of the Chinese Academy of Sciences and Chinese Academy of Engineering, said that the design of the new hall was majestic and frugal and showed the disaster of the Chinese in a very realistic methodology".

03

04

05

04. 史料陈列厅出口
05. 史料陈列厅入口

对页
06. “和平之舟”船头侧面“化剑为犁”的立面造型。
07. 祭场，黑色的花岗岩石碑前，燃烧着遇难者的长明火，人们通过献花圈、敬香、默哀等形式，寄托对遇难者的哀思和反对战争、祈祷和平的愿望。
08. 和平公园

P162-163 **页**
09. “万人坑”遗址内景
10. 胜利之墙，著名雕塑家吴为山教授主持创作设计，墙上的浮雕以象征胜利的 V 字造型向两边延展。
11. 冥思厅（图为纪念馆冥思厅中，陈列着由李长春和李源潮共同题写的“六个让、两个把”，发人深思）

04. Exit of Historical Material Exhibition Hall
05. Entrance of Historical Material Exhibition Hall

Opposite page
06. The pattern on the side of the ‘the boat of peace’ means ‘change the sword into plowshares’
07. Mourning Square, the everbright fire of the victims burning in front the black granite monument. People come here to lay the wreaths. Present the incense sticks and stand in silent tribute, by which they show their grief to the victims and the wish of world peace.
08. Peach Park

P162-163
09. Interior of the “Pits of Thousands of Bodies” Site
10. Victory Wall, designed and created by “famous sculptor Wu Weishan, The relief of letter V extends to both sides, which stands for the victory.
11.Meditation Hall(Thought provoking inscription by Li Changchun and Li Yuanchao displayed in Meditation Hall of the memorial)

06

07

08

09

10

让白骨可以入睡
让冤魂能够安眠
把屠刀化铸警钟
把逝名刻作史鉴
让孩童不再恐惧
让母亲不再泣叹
让战争远离人类
让和平洒满人间
Let the white bones fall into sleep
Let the grievance souls rest peacefully
Turn the slaughtering sword to the cast of a warning bell
Turn the names of the departed to be engraved in history
Let the children no longer face with the fear
Let the mother no longer shed their tears
Let war be distant from all mankind
Let peace be spread fully to the world

11

Suzhou Museum

苏州博物馆

苏州博物馆成立于1960年，馆址假借于1860年始建的全国重点文物保护单位太平天国忠王府。1999年，为适应苏州经济的快速发展和广大市民休闲益智的文化需求，苏州市决定建造新的苏州博物馆。2006年，由世界著名建筑大师贝聿铭担纲设计的苏州博物馆新馆在金秋时节向世人露出崭新的容颜。

苏州博物馆新馆位于历史保护街区东北街和齐门路交会处，紧临世界文化遗产拙政园和全国重点文物保护单位太平天国忠王府，建筑面积逾19 000m^2，加上修葺一新的忠王府，总建筑面积达26 500 m^2。

新馆的设计结合了传统的苏州建筑风格，把博物馆置于院落之间，使建筑物与周围环境相协调，黑白灰的主色调与苏州传统的城市肌理相融合。中轴线对称的东、中、西三路布局与忠王府格局相互映衬，十分和谐。中心部分是入口处、大厅和主庭院；西部为展区；东部为现代艺术画廊、教育活动区、茶水服务区以及行政管理功能区等，该部分还将成为与忠王府连接的实际通道。八角形大厅是通过对传统的苏州建筑和中国建筑要素的几何形状转变以及重新诠释设计出来的，它是所有参观者的导向并为去博物馆所有展区提供通道。

Founded in 1960, Suzhou Museum took over the site of Zhong Wang Fu, which was built in 1860, a unit key cultural relic protection of state level. In 1999, the City decided to build a new museum to meet the citizens' cultural demands , as well as complement the rapid growth in economy. The new Suzhou Museum, designed by the worldly renowned architect, Mr. I. M. Pei, was opened to the public in the autumn of 2006.

New Suzhou Museum is located at the intersection of Dongbei Street and Qimen Road. These two streets are also designated as historic streets, adjacent to Zhong Wang Fu and the Humble Administrator's Garden. The Museum covers an area of over 19,000 square meters. Including the recently renovated Zhong Wang Fu, it is then 26,500 square meters in total.

All the traditional Suzhou architecture styles are carefully blended in one design. The main building is surrounded by the courtyards, which stands in such a harmony with all other buildings. Black, white and grey are the dominant colors throughout the museum buildings, reflecting the traditional Suzhou colors.

The new museum is divided into three major parts: the central path, east and west wings, which is consistent with the design of Zhong Wang Fu. The entrance, lobby and garden are all parts of the central path, galleries with the permanent collection are located on the west. The modern art galleries, tea house, library, and offices are on the east. The east wing is also connected to Zhong Wang Fu. The octagonal lobby represents a transformation and interpretation

01

02-03. 平面

对页
01. 博物馆全景俯瞰

02-03.ground-model

Opposite page
01. Aerial view of the museum

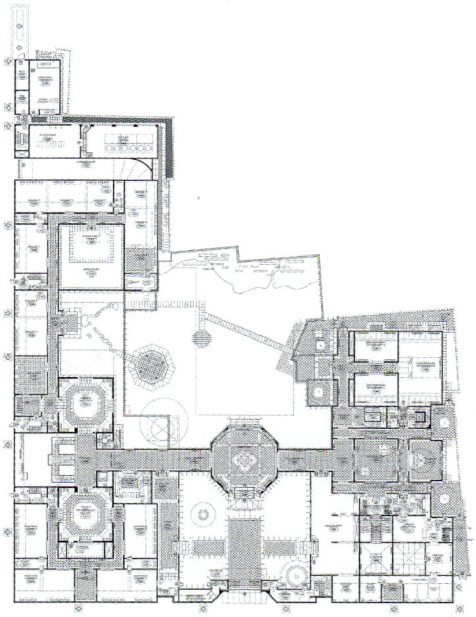

02

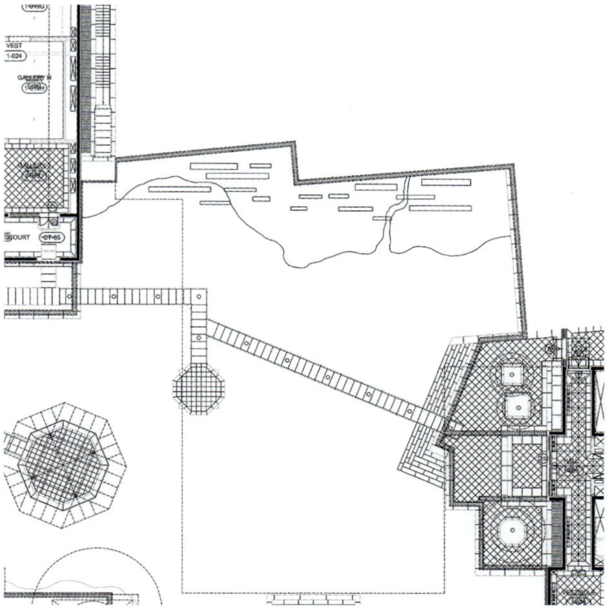

03

01

04

参照国际通行惯例，临时性的展区设在人流路线的入口处。除了字画、双塔瑰宝、明清瓷器和苏州工艺美术品展区外，新馆还布置了特色家具展区，并在人流路线的终点处，用传统手法复原了一间宋代民间厅堂，展示当年的工艺品和家具。字画展区设在自然采光的八角形大厅的二楼，高高的墙面为展示字画长卷提供了背景上的便利。现代美术作品设在主庭院东边的一个特别展区里。由于地块的大小、高度的限制以及博物馆设计规划的要求，相当一部分的博物馆功能空间安排在地下室。游人可以通过室内荷花池上方的悬臂楼梯到达地下临展厅。

of the geometrical design of the traditional Suzhou as well as Chinese architecture. It provides the visitors with an easy access to all galleries in the museum.

To conform with the international convention, the temporary galleries are thus located at the beginning of the visiting route. Besides the pagoda, porcelain, art and craft galleries, the new museum arranges a furniture display of Ming Dynasty. At the end of the tour, a very noticeable pavilion of Song Dynasty was restored based on the traditional way of construction. On the second floor, two tall galleries facing each other were designed to primarily display paintings and calligraphy. In addition, the temporary galleries, audio and visual room, auditorium, restrooms, offices, storage vaults and machinery room are all located in the basement level.

04. 主庭院

对页
05. 创意山水

04. Main court yard

Opposite page
05. Creative Landscape

05

06

如何在如此厚重的历史文化氛围中造一座走向 21 世纪的建筑，是一个世界性的难题。可是苏州博物馆新馆，人、建筑、环境的理念在这里得到了极高境界的融合，置身馆内，仿佛置身画中。

It certainly presents a challenge how we should leap into the 21th century with a museum filled with history and traditions. However, with the successful merging of modern architecture, human and wonderful environment, Suzhou Museum undoubtedly has accomplished a platform for you and me to enjoy, as if we were part of a beautiful painting.

06.“陶冶之珍”展厅
07.“吴门书画”展厅

对页
08. 精致的门庭

06. Porcelain Exhibition Hall
07. Wu Painting & Galleries

Opposite page
08. Exquisite lobby

07

Nantong Museum

南通博物苑

南通博物苑是近代著名的爱国实业家、教育家张謇于1905年（清光绪三十一年）创建的，位于江苏省南通市风光秀丽的东南濠河之滨，占地面积72 000 m²，是中国人创办的第一座博物馆。

早年的博物苑有中馆、南馆、北馆等主要建筑，用于陈列天产、历史、美术、教育四部标本、文物；以后逐年建造了藤东水榭、味雪斋、相禽阁、国秀亭、假山、水池等园林建筑设施，在北馆和中馆之间中轴线的两侧，还对称设置了荷花池、国秀坛、风车和水塔，至今基本格局未变。1981年张謇故居濠南别业划归博物苑管理，经全面修缮，恢复原貌，供人们观光瞻仰。

南通博物苑各馆舍，中西合璧，风格多样，园林部分亦中西兼备，各具特色，具有典型的清末民初建筑特征。其中濠南别业由张謇先生亲自培养的南通籍中国第一代建筑师孙支夏设计建造。这座具有欧洲古典风格的建筑，作为吸收西方建筑艺术的典范和近代优秀建筑，分别载入《中国建筑史》和《20世纪中国建筑》。

Nantong Museum was established by Zhang Jian, the famous patriotic industrialist and educationist in Modern China, in 1905(Qing Dynasty, Kuang-hsu 31st year). As the first public museum founded by Chinese, it is located alongside the scenic Southeast Haohe River and covers an area of 72,000 square meters.

The early main buildings (Middle Hall, South Hall and North Hall, etc.) were used to display the cultural relics, paintings and calligraphic works, and natural science specimens. In subsequent years, some landscape buildings (Tengdong Waterside Pavilion, Weixue Room, Xiangqin Pavilion, Guoxiu Kiosk, rockeries and pools, etc.) were built. As the basic pattern that has not changed, on both sides of the axis between North Hall and Middle Hall, the lotus ponds, Guoxiu Altars, windmills and water towers were set up symmetrically. In 1981, the former residence of Zhang Jian, Haonan Villa, was managed by Nantong Museum, which was renovated completely and reinstated for tourism.

Every construction in Nantogn Museum, is a combination of Chinese and Western styles, and has the typical architectural features of the Late Qing Dynasty. Specially, the Haonan Villa was designed by Sun Zhixia who was one of the first generation of architects of China and cultivated by Zhang Jian in Nantong. With the European classical style of architecture, this construction was written into *History of Chinese Architecture* and *20th Century Architecture* as a model of absorption of western art and early modern architecture.

01

02

03

04

南通博物苑融中国古代园囿与近代博物馆于一体，将民族特色与科学内容相结合，是一座“园馆一体”的城市园林式综合性博物馆，在中国博物馆发展史上具有划时代的意义。

1988 年南通博物苑被国务院公布为全国重点文物保护单位。

With the combination of modern museum and Chinese ancient garden styles and the combination of national characteristics and scientific content, Nantong Museum is a local comprehensive museum in landscape style. It has epoch-making significance in the Chinese museum history.

In 1988, Nantong Museum was approved by the State of Council as a national major historical and cultural site under state protection.

02. 博物苑北馆旧影
03. 博物苑东馆旧影
04. 博物苑谦亭旧影

对页
01. 博物苑新馆全景

02. Old photograph of North Hall
03. Old photograph of East Hall
04. Old photograph of the Pavilion

Opposite page
01. Panorama of the museum

05

06

07-08

07. 濠南别业
08. 濠南别业

对页
05. 中馆和南馆
06. 中馆和南馆旧影

07. Old photograph of Haonan Villia
08. Haonan Villia

Opposite page
05. Middle Hall and South Hall
06. Old photograph of Middle Hall and South Hall

09

10

11

10-11. 细部

对页
09. 细部

10-11. Architectural details

Opposite page
09. Architectural details

Yangzhou Museum (China Block Printing Museum)

扬州博物馆（中国雕版印刷博物馆）

扬州双博馆是扬州中国雕版印刷博物馆和扬州博物馆的合称。建筑的设计体现了历史与现代、艺术与自然的交融合一。建筑外墙设计成完全的曲线，暗含扬州是一座运河城市，同时因紧邻明月湖，曲线自然合围出两片形态优美的“荷叶”，与湖面及沿湖建筑相融合、呼应，形成湖光山色于一体的写意山水建筑群。石材外墙为土红色，是对扬州英雄城市和汉唐繁盛的表达，体现了建筑师挖掘扬州文化内涵特质的匠心独运。大厅的外墙采用颇具现代感的抓点式玻璃幕墙构造，两馆的天花中央使用大面积的玻璃封顶，将阳光直接引入大厅，室内影随光动，增强了室内的动感流线和建筑的现代感。

扬州双博馆的建筑面积达 25 000 m^2，总高 22.9 m，建筑主体三层，分为两区，东区为扬州中国雕版印刷博物馆，西区为扬州博物馆。中间共享大厅将两馆有机联结，两馆在平面上各成一区，在立面造型上融为一体，展览、库房、研究、办公、餐饮休闲等功能区域各成系统而有机联系。建筑设计展览面积 9 319.4 m^2，设有 7 个基本陈列展厅和一个临时展厅。

That is Yangzhou's new cultural landmark-Yangzhou Double Museum. Yangzhou Double Museum is the combination name of China Block Printing Museum at Yangzhou and Yangzhou Museum. The architectural design presents the combination of historical and modern city, art and nature. The architecture's external wall is designed as a whole curve, which indicates that Yangzhou is a city with canal. Meanwhile, as it is close to Bright Moon Lake, the curve looks like two graceful lotus leaves, reflects with the lake and architecture along the lake, forms a landscape architectural group. The color used for the stone material of the external wall is soil red, which expresses the heroical Yangzhou city and the flourish in Han and Tang Dynasties, presents the architect's inventive mind of digging Yangzhou's cultural connotation. The external wall of the hall uses modern glass curtain wall structure; it uses large-area glass covers in the middle of the two museums' ceiling. By this way, the sunshine can go directly into the hall, shadows move with sunlight, which enhance the inside flow line and architecture's modern sense.

Yangzhou Double Museum covers an area of 25,000 square meters, with total height of 22.9 meters. The main building has three floors and is divided into two districts. The east district is China Block Printing Museum at Yangzhou and the west district is Yangzhou Museum. The hall in the middle combines the two museums together, and makes them separate in the plane but still a complete piece in façade configuration. The functional areas such as exhibition, warehouse, research, office, and dinning & leisure have their own systems but still connect with each other. The exhibition building area is 9,319.4 square meters, with seven basic and one temporary exhibition halls.

01

01. 博物馆全景
02. 大厅

01. Panorama of the museum
02. Hall

02

03. 二楼西区共享空间
04. 大厅

03. Shared space of western zone on 2nd floor
04. The hall

对页
05. 史可法纪念馆，1951 年 11 月至 1988 年 1 月为扬州博物馆所在地。
06. 天宁寺，现为扬州中国佛教文化博物馆，1988 年 2 月至 2005 年 9 月为扬州博物馆所在地。

Opposite page
05. Shi Kefa Memorial, Location of Yangzhou Museum from Nove.1951 to Jan.1988
06. Tian Ning Temple, China Buddhist Culture Museum, was once the location of Yangzhou Museum from Feb.1988 to Sep.2005.

广场作为主体建筑的铺叙和延伸，是建筑设计理念的延续。不断跃升的主立面广场平台象征着历史发展曲折前行。从 2 500 年建城史中提炼出的 10 位历史人物雕像点缀在广场各处，给人以无尽的遐思，营造了独特的博物馆氛围。

As the spread and extension of the main architecture, the square is the extension of architecture design idea. The continuous ascending main elevation square platform symbolizes the tortuous development of history. Ten sculptures of historical people who were selected from 2,500 years' city history stand in different locations of the square, give people infinite thinking and also build a unique museum atmosphere.

05

06

Changzhou Museum

常州博物馆

常州博物馆创建于1958年，是一座地方综合性博物馆。2005年3月动工建设新馆，2007年4月28日正式落成对外开放。新馆位于常州市新北区龙城大道1288号，博物馆与规划馆两馆合一，位于市民广场西侧，与常州市行政中心、大剧院、奥体中心一体规划。博物馆周围设有专门的绿化带以及水体等缓冲带，形成优美的天际轮廓线。

Established in 1958, Changzhou Museum, is a comprehensive regional museum. The new museum complex started its construction in Mar., 2005 and finally opened in Apr. 27th, 2007. It stands on No. 1288, Longcheng Road, Xinbei District in Changzhou, including the museum and the urban planning exhibition hall. It was planned to be placed to the west of the citizen square, together with the administration center, theater and the Olympic sports center. The museum is surrounded by special green belts and water buffering zone, which also creates an elegant skyline.

01

常州博物馆建筑设计方案，立意为“传统与现代的对话”。建筑体量巨大，空间互有穿插，并形成与广场的呼应和向心感。建筑北侧的博物馆与南侧的规划馆及两馆之间的共享大厅三部分通过屋面大型飘板有机结合成为一个弧形整体。建筑采用了较多的玻璃幕墙和石质材料，具有较强的现代建筑特征。

The architecture design plan means the dialogue between tradition and modernness. The museum has big volume and responses with the square to create and increase the centrality. The building complex is divided into three parts: the museum in the north, the urban planning exhibition hall in the south and the communion hall in-between, all together organized by a huge floating slab on the roof. Meanwhile, the extensive use of glass curtain wall and stone material also endows modernity to the museum.

01. 清凉寺，1983年至2007年3月为常州博物馆所在地。

对页

02. 博物馆正门

03. 博物馆外景

01. Qingliang Temple, location of Changzhou Museum from 1983 to March 2007

Opposite page

02. Front gate of the museum

03. Exterior of the museum

两馆占地24791 m²，建筑面积34930 m²，总投资2.8亿元。其中博物馆建筑面积为19863 m²，地下一层，地面4层，局部三层，展区面积约10000 m²，库房、设备用房约4303 m²，办公用房约5560 m²。另有3232 m²的公共通道、

The site area of the complex is 24,791 square meters, and the total area is 34,930 square meters. The whole project consumed 280 million yuan. The total area of the museum is 19,863 square meters, one floor underground, four floors aboveground (partly three floors). The exhibition area is 10,000 square meters, including storerooms and

02

03

学术报告厅和贵宾接待室为两馆所共用。整个建筑外观新颖、环境幽雅、设施齐全、功能先进，富有时代气息。常州博物馆正以崭新的面貌，热忱欢迎国内外宾客参观游览。

facility rooms (approximately 4,303 square meters), the office rooms (5,560 square meters). Besides, there are some spaces shared by the two buildings, including the public inner channel, academic report halls and VIP reception rooms (3,232 square meters).

The museum is a perfect combination of novelty in appearance, elegance in environment, completeness in facilities and advantages in functions, and also shows the characteristic of the times. Now, Changzhou Museum is waiting for visitors at home and abroad cordially with its new visage.

04

05

04. 常州博物馆外围绿化带及水体缓冲带
05. 展陈区

对页
06. 展陈区

04. Outer green belt and water buffer belt
05. Exhibition area

Opposite page
06. Exhibition area

Wuxi Museum

无锡博物院

无锡北依长江、南临太湖，京杭大运河穿城而过，贯通南北。早在 3 000 多年前，泰伯就带领吴地先民开凿运河，引水造田。无锡博物院的造型厚重而富有水的灵性，建筑设计师用水性的刚柔，雕塑般的语言，成功体现了“水光石色”的理念，并隐喻了吴地先民对水的依恋和崇拜，构思精妙，创意新颖。

无锡博物院成立于 2007 年 10 月 15 日，由原无锡博物馆、无锡革命陈列馆和无锡科技馆“三馆合一”组建而成，位于无锡城市客厅——太湖广场，是无锡地区最大的公共文化服务设施。项目东西跨度 260 m，南北宽 78 m，高 42 m，建筑面积 71 000 m^2。整个建筑采用现代化全钢架结构，是无锡的标志性建筑，也是无锡建筑史上的一座里程碑。

无锡博物院遵循“以人为本、观众至上”的设计理念，统一布局，形成“文博历史、科技之光、影视教育、专题特色”四大板块。馆内有 23 个常设展览，并相继建成无锡徐悲鸿艺术展馆、集成电路体验馆和虚拟

Wuxi , the star city built on the shore of Taihu Lake, is a major spot in the Lower Yangtze Delta. Crisscrossed with old canals, the city enjoys a long history of prosperity brought by water. The canalization of this area could be dated back to 3000 years ago, when the fugitive princes of Zhou Dynasty, Taibo began to develop the local agriculture and waterways with his people. The modern all-steel structure was adopted for the construction, therefore, the whole building appears thick and stable, the lines of elevation exhibit hardness together with softness, representing the spirit of Water Culture in Wu Area, and the design concept is well expressed.

Wuxi Museum, founded on October 15th, 2007, composed of the former Wuxi Revolutionary Exhibition Hall, former Wuxi Museum and former Wuxi Science and Technology Museum, located in Wuxi living room—Taihu Lake Square, is the largest facilities provider for public cultural service in the city. The new museum, 260 meters long from east to west, 78 meters wide from north to south, 42 meters high, covering the area of 71,000 square meters and adopting modern all-steel structure, has become the landmark architecture of the city as well as a milestone in the history of the architecture in Wuxi.

Wuxi Museum adopts the design concept of respecting humanity and audience,, and has formed the Four Plates, i.e., Literature and History, The Light of Science and Technology, Film and TV Education and

01

01. 夜景

对页
02. 一层平面图
03. 剖面
04. 博物院外景

01. Nightscape

Opposite page
02. Plan of the ground floor
03. Section
04. Exterior of the museum

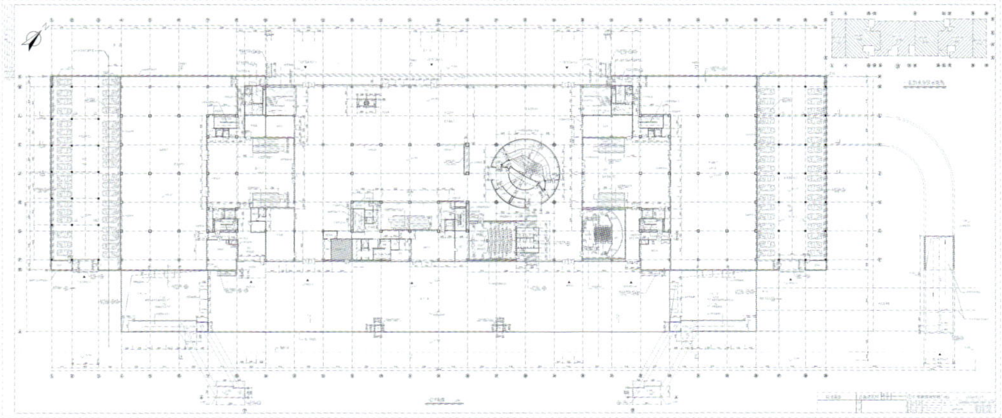

03

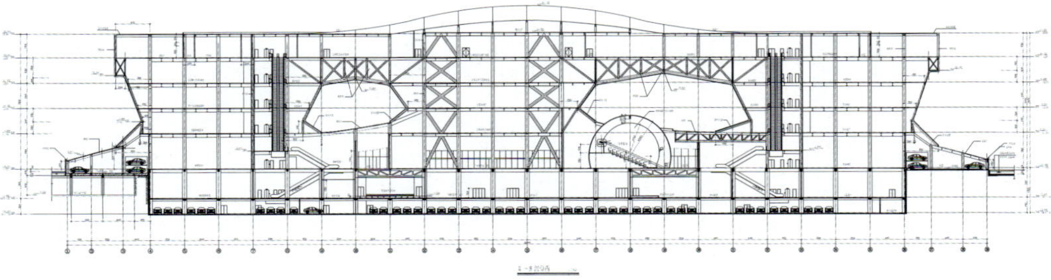

04

02

西方艺术馆，同时还建有世界先进的特效影视剧场（球幕、4D、动感影院），充分反映了无锡地域文化特色，彰显了无锡城市独特的个性魅力，是一座集教育引导、观赏娱乐、互动体验为一体的多功能综合性现代化的大型博物馆。

Specialty and twenty-three routine exhibitions. Additionally, subsidiary Xu Beihong Art Gallery, Integrated Circuit Experience Hall, Virtual Gallery of Western Art and the advanced theaters in the world (including 4D theatre, Dome theatre and Simulation theatre) were established gradually, which not only represent the historical and local characteristics of Wuxi but display the unique charm of Wuxi. Therefore, Wuxi Museum is a modern large and comprehensive museum with the multifunction of education, entertainment and interaction.

05

06

05. 历史基本陈列，无锡城市故事
06. 革命史展厅

对页
07. 科技展厅
08. 书画展厅

05. History Exhibition of Wuxi City
06. Revolution History Exhibition Hall

Opposite page
07. Technology Exhibition Hall
08. Calligraphy Exhibition Hall

07

08

Zhejiang Provincial Museum

浙江省博物馆

01

02

浙江省博物馆成立于 1929 年 11 月，其前身为浙江省西湖博物馆，是我国早期建立的有影响的博物馆之一。除馆藏文物和展览在国内外有较高声誉外，博物馆的建筑富有现代气息和江南仿古建筑特色，其所处环境更为世界一流。2009 年起，浙江省博物馆形成孤山馆区和武林馆区两大馆区格局，成为仿古与现代结合的博物馆建筑风格。

孤山馆区位于杭州西湖孤山南麓，左邻平湖秋月，右接中山公园，面临西湖，风光如画，景致宜人。早期馆舍以清代和民国建筑为主，由王阳明祠、文澜阁、忠烈祠、圣因寺罗汉堂、太乙分清室等数处合并而成。1993 年改扩建工程竣工，扩建的新馆建筑面积约 8 000 m²，陈列面积约 5 000 m²。

Established in November 1929 with the original name of Zhejiang Provincial West Lake Museum, Zhejiang Provincial Museum(ZPM for short) is one of the influential forerunning museums in China. Apart from enjoying a reputation for quality collections and exhibitions at home and abroad, ZPM is also renowned for its modern and pseudo-classic Jiangnan architecture and the picturesque surroundings. In 2009, ZPM formed a pattern of two museums–Gushan Branch and Wulin Branch, thus an architecture merging pseudo-classic and modern styles coming into being.

Since its establishment, Gushan Branch, situated at the southern foot of Gushan (the Solitary Hill), adjacent to the Autumn Moon over the Calm Lake on the left and Zhongshan Park on the right, has enjoyed a favorable and charming landscape facing the West Lake. In its early stage, the architecture, merging Wang Yangming Ancestral Temple, Wenlan Pavilion, Zhonglie Ancestral Temple,

01.20 世纪 30 年代浙江省立西湖博物馆前门全景
02. 博物馆旧影（摄于 20 世纪 50-70 年代）

对页
03. 孤山馆区曲线连廊
04. 孤山馆区主楼建筑

01. Overall view of the Zhejiang Provincial West Lake Museum in the 1930s
02. Historical photo (1950s-1970s)

Opposite page
03. Zigzag corridor at Gushan Branch
04. Main building of Gushan Branch

03

04

文澜阁古建区与改扩建后新馆区融为一体，形成了富有江南园林特色的建筑群，有大小15个展厅。整个建筑群按照“园中馆，馆中园”的构思设计，由民族形式的大屋顶单体建筑和曲线连廊组合而成，总体色调以米黄色为主。1999年底，在孤山馆区东侧又新建浙江西湖美术馆，建筑面积约2500 m^2，陈列面积约1500 m^2，成为浙江省博物馆重要的艺术展览和交流场所。

武林馆区于2009年12月22日正式对外开放。其地处市中心，位于杭州运河之畔的西湖文化广场E区，总建筑面积20991 m^2，陈列面积约7600 m^2，室内设地下一层和地上三层。整个建筑设计以运河文化为背景，结合现代文明的意象，体现秀外慧中的特色。建筑采用幕墙玻璃，空间高、采光好，适合举办各种大型展览。

Arhat Hall of Shengyin Temple, Tai Yi Fen Qing Shi, was mostly of the style in the Qing Dynasty and the Republic of China. In 1993, the Branch was expanded to an area of about 8,000 square meters, and an exhibition area of 5,000 square meters.

Gushan Branch, with the perfect combination of Wenlan Pavilion and the expanded architecture of 15 big and small exhibition halls, has formed an architectural complex featured with Jiangnan garden style. The complex was imbued with the idea of "garden in the museum and museum in the garden", integrating the overhanging-roof Chinese monomer construction with the curved corridor against a cream-colored background. At the end of 1999, adjacent to the Gushan Branch on the east was erected Zhejiang West Lake Gallery, with a floor area of 2,500 square meters and an exhibition area of 1,500 square meters, and it has now become a key center for cultural exhibitions and exchanges of the ZPM.

Wulin Branch, situated in the E zone of West Lake Cultural Plaza on the Grand Canal, the downtown area of Hangzhou, was open to the public on 22nd, December, 2009. With a floor area of 20,991 square meters and an exhibition area of 7,600 square meters, the Branch has one floor underground and three floors above. Based on by the Grand Canal Culture, the whole architecture design, with beauty externally and knowledge internally, is a combination of modern civilization images. The architecture, adopting glass curtain walls, has expansive space and good lighting, suitable for various large-scale exhibitions.

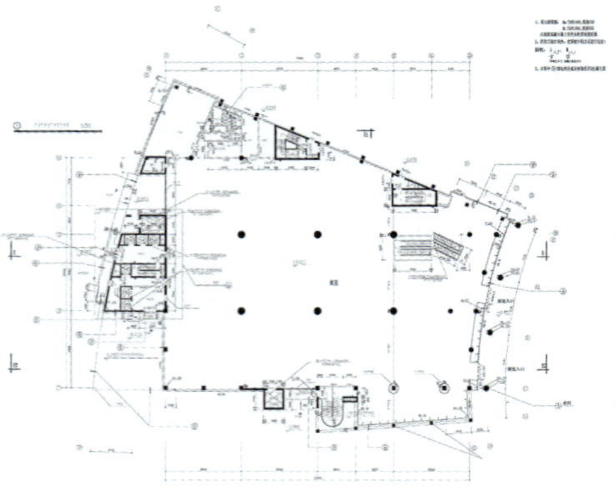

05

06

07

07. 武林馆区一楼大厅　　07.Lobby of Wulin Branch

对页　　**Opposite page**

05. 武林馆区首层平面　　05.Plan of the ground floor in the Wulin Branch

06. 武林馆区　　06.Wulin Branch

China National Silk Museum

中国丝绸博物馆

位于杭州西子湖畔玉皇山下的中国丝绸博物馆是国家级的专业博物馆，占地面积 50 000 m^2，建筑面积 8 000 m^2，总体布局灵动飘逸，建筑主体造型新颖独特。富有江南水乡特色的马头墙围合的大厅，其穹窿形式的玻璃中庭，犹如一颗蚕茧，象征着建筑物的鲜明个性，成为杭州市西湖黄金旅游线上的一颗闪闪发光的璀璨明珠。

博物馆于 1992 年 2 月正式对外开放，2004 年 1 月 1 日起对公众实行免费开放。它是中国最大的纺织品文物收藏、展示、保护、研究和教育中心，是中国丝绸文化的宣传窗口。博物馆环境清静幽美，拥有桑园花草、亭台楼阁、小桥流水。

展馆于 2004 年完成了基本陈列和馆区环境的整体改造。基本陈列的主要内容为丝绸故事和丝绸工艺。"中国丝绸的故事"分为丝绸厅、丝绸之路、服饰厅，"丝绸工艺陈列"分为蚕桑厅、染织厅、织造坊。馆内还设有临展厅，不定期举办各种特色展览。

Seated at the foot of Yuhuang Hill which lies at the bank of West Lake, China National Silk Museum, a national specific museum that covers an area of 50,000 square meters and has a floor area of 8,000 square meters, has a flexibly-arranged plan and a uniquely-designed main building. The entrance hall is surrounded by a typical kind of wall called Matou Wall with Jiangnan (literally, south of the Yangtze River) features and covered by a cocoon-shaped glass dome, which represents the unique characteristic of the museum, is becoming the most resplendent spot in the golden tour line of West Lake.

The museum was formally opened in February 1992 and free to the public since January 1, 2004. It is the largest center for textile heritage collection, exhibition, conservation, research and education in China; thus becomes the communication window for Chinese silk culture. Also, along with the typical Chinese gardens, pavilions, bridges and streams, the museum itself is a quiet and beautiful place to stay.

The overall transformation of the environment and the basic display and exhibition areas was completed in 2004. The stories, arts and crafts of silk are now the main contents in the basic display area of the museum. " The stories of Chinese Silk" is displayed by three smaller exhibition halls named " The hall of silk", " The Silk Road" and " Traditional Clothes". " The Arts and Crafts of Silk" is showed

01

01. 蚕农家园场景

对页
02. 博物馆俯视图
03. 博物馆外景

01. Scence of silkworms farmers' home

Opposite page
02. Top-view of the museum
03. Exterior

02

03

通过展览，博物馆向各国人民展示了凝聚着中国人民智慧和世界人民友好交往的实证，让他们从历史的延续中加深对悠久、灿烂的中华文明的认识和了解，感受丝绸人文精神的内涵和魅力。

in "The hall of Sericulture", "The Hall of Dyeing" and "The Hall of weaving". In the temporary exhibition hall of the museum, various distinctive exhibitions are held from time to time.

Through exhibitions, the museum shows people world-wild the demonstrations which represent the friendship between Chinese people and the world from ancient times. Visitors will no doubt have more profound understandings of the long, brilliant Chinese culture and be deeply touched by the connotation and enchantment of the silk culture, which is also the culture of Chinese people.

04

05

04. 序厅
05. 文物厅

对页
06. 染织厅
07. 织造厅

04. Prelude hall
05. Relics hall

Opposite page
06. Coloring hall
07. Weaving hall

06

07

Wenzhou Museum

温州博物馆

温州博物馆是一所综合性地方博物馆，始建于 1958 年，原址坐落在江心屿，馆舍系由清代古刹兴庆寺改建而成，占地面积 1 800 m^2。在市委、市政府的关心和支持下，2003 年 10 月，位于世纪广场西侧的新馆落成并正式对外开放，一座崭新的现代化、多功能的博物馆屹立在浙南大地上，成为“文化温州”的一道亮丽风景。

温州博物馆建筑设计灵感源自雁荡山的奇峰翠峦，展现了追求雄浑、朴茂的艺术神韵。大门照壁饰以大榕树石雕，寓意温州人勤劳务实、百折不挠的开拓创新精神。中厅花岗岩壁面悬挂盘古开天、女娲补天、夸父追日、精卫填海、后羿追日、嫦娥奔月、神农伏羲、燧人取火、大禹治水等九幅中国传统神话题材的铜雕壁画，力图表现中华文明的博大精深。

本馆建筑面积约 26 000 m^2，馆内展区总面积 8 000 m^2。历史馆建筑面积面积 2 100 m^2，展示“温州人” 5 000 年来的历史足迹。展区另设书画馆、陶瓷馆、自然馆、工艺馆等六个专题陈列。展示内容包括陶

Wenzhou Museum is a local comprehensive museum, originally located on the Jiangxin Island in 1958, rebuilt on the base of the old Xingqing Temple of Qing Dynasty, covering an area of 1,800 square meters. Under the concern of Municipal government, the new museum situated on the west of Century Square was completed and opened to the public, in October, 2003. From then on, a brand new, modern and multi-functional museum has been standing on the South of Zhejiang Province, becoming a unique urban sight.

The design concept of Wenzhou Museum stemmed from the unique views of Yandang Mountain, which displays a vigorous and historical feeling. The main doors are decorated with the stone carving of great banyan, which shows the spirit of diligence, pragmatism and perseverance of the people of Wenzhou. On the granite walls of the hall hang nine frescos which are based on the traditional fairy tales of China(*Pangu Kai Tian, Nüwa Bu Tian, Kuafu Zhui Ri, Jingwei Tian Hai, Houyi She Ri, Chang'e Ben Yue, Shennong Fuxi, Suiren Qu Huo, Dayu Zhi Shui*), which manage to display the profound traditional culture of China.

The museum has a floor area of 26,000 square meters, in which the inner exhibition halls cover 8,000 square meters. Thereinto, historical exhibition hall space uses 2,100 square meters, showing the Wenzhou people' s history. The exhibition halls cover twenty major categories

01

01. 北宋阿育王漆塔

对页
02. 博物馆天井
03. 博物馆外景

01. Asoka lacquer pagoda in North Song Dynasty

Opposite page
02. Patio of the museum
03. Interior of the museum

瓷器、青铜器、彩塑、砖雕、漆器、书画等20个门类，而尤以陶瓷器、彩塑、书画为特色。展览充分运用空间组合，力求在内容设计和艺术形式有所突破和创新，使一大批文化瑰宝得到充分的展示，文物内涵得到更深的发掘，从而使更多观众热爱传统文化，熟悉温州历史。

of Chinese Art , such as Ancient Ceramics, Ancient Bronze, Ancient Sculptures, Brick Carvings, Lacquer-work Painting and Calligraphy, in which the Ancient Ceramics, Ancient Bronze, Painting and Calligraphy are prominent. The exhibition makes full use of the spatial array, striving to make a breakthrough both in the field of design content and art form. A host of cultural treasures displayed to the public and the meanings of historical relics discovered will make an increasing number of visitors admire traditional culture and familiar with the history of Wenzhou.

02

03

04

05

04. 博物馆序厅
05. 书画馆场景

对页
06. 历史馆——场景之首饰龙
07. 陶瓷馆场景

04. Lobby of the museum
05. Calligraphy Exhibition Hall

Opposite page
06.Historical Exhibition Hall — scene of Jewelry Dragon
07. Ancient Ceramics Exhibition Hall

06

07

Anhui Museum

安徽省博物馆

安徽省博物馆于 1955 年开工建设，1956 年 11 月 14 日正式对外开放，是新中国成立后开工建设的第一批省级综合博物馆之一。毛泽东、朱德、刘少奇、周恩来、邓小平、江泽民等党和国家领导人先后来此视察。陈列展览大楼见证了新中国博物馆事业的变迁，成为中国博物馆发展史上具有标志性意义的历史建筑。陈列展览大楼 2007 年被安徽省人民政府公布为“省级文物保护单位”。

陈列展览大楼面南背北，建筑面积 11 580 m^2，展览面积 7 826 m^2。大楼整体呈“П”形，平面规矩，构图强调中轴对称，立面造型为渐次升高的形体，内部具有高空间、小库房等特点，保留了典型苏式展览建筑的浓重印记。主楼高耸挺拔，回廊宽缓伸展。出于经济和工期因素的考虑，大楼不设尖塔和柱廊，去除了烦琐的装饰。建筑立面处理简洁，室内外无过多装饰纹样，中部略微凹下，突出了下部的入口，使上部的塔楼造型线条更加挺拔、雄伟。整体形象明快，具有很强的展览建筑特色。

Anhui Museum started construction in 1955 and was open to public on November, 14th, 1956. It is one of the earliest province-level museums built after the People's Republic of China was founded. It has proudly presented itself to several leaders of our party and the country, including Mao Zedong, Zhu De, Liu Shaoqi, Zhou Enlai, Deng Xiaoping and Jiang Zemin. The Exhibition Building witnesses the development of Chinese museums, and has become a symbolic building in Chinese museums' history. It was listed as a "Heritage Site under Province Protection" in 2007 by the People's Government of Anhui Province.

The Exhibition Building faceing south and backing north, has a floor area of 11,580 square meters and an exhibition area of 7,826 square meters. The building is in a П shape and emphasizes symmetry in its planar design, while vertically the height is increasing gradually. The inside design features tall ceilings and small storerooms, which are basically from traditional Russian-style exhibition structure. The main building is a tall and straight one, accompanied by wide and loosely plotted corridors. Neither towers or colonnades were built because of economical and time considerations, also redundant ornaments were avoided. The whole design is basically simple but practical, without extra decorations. With the middle part slightly concave, the lower entrance is highlighted, which accentuates the tower shape of the upper part. The overall look is lucid and lively, with typical exhibition characteristics.

01

01. 建馆初期的陈列展览大楼

对页
02. 博物馆外景
03. 开阔敞亮的中厅

01. Exhibition Building in the early days of the founding of the museum

Opposite page
02. Exterior of the museum
03. The spacious bright atrium

02

03

1968 年大楼主楼建筑正立面进行局部调整，增建观礼平台，并在主楼 4 层增建挡墙，下方加立 6 根支撑立柱，柱头分别有“韶山毛泽东故居”、“遵义会议旧址”等 10 幅浮雕，客观反映了安徽省博物馆被各个历史时期所赋予的时代价值。

与北京展览馆等苏式建筑相比，安徽省博物馆在保留苏式建筑韵味基础上，又有了新的探索。

In 1968 the main building was partially reconstructed, a viewing platform was added, also a wall was built on the 4th floor with 6 supporting pillars, decorated by 10 embossments including "The former residence of Mao Zedong in Shaoshan", "The Zunyi Conference Site". These embossments represented the important positions of this museum during different historical periods.

Compared with other Russian-style constructions, like Beijing Exhibition Center, Anhui Museum exhibits valuable explorations while keeping the essence of the Russian-style building.

Fujian Museum

福建博物院

从福建省立科学馆到福建省人民科学馆，从福建省人民科学馆到福建博物院，最后再到福建博物院新馆的建成，近 70 年峥嵘岁月中，福建博物院根据社会发展对文化的需要，改善服务环境设施；结合福建省经济发展的契机，创新服务理念，致力于将自身建设成为提供全方位文化服务的“开放性博物院”。

主体建筑呈现福建风格，蕴涵和谐。新馆于 2002 年 10 月建成，占地面积 60 000m²，建筑面积 36 000m²。主体建筑是数个具有福建文化特色元素的集合：富有福建民居特色的“几”字形屋顶，体现“闽”字内涵的图腾柱，汇集中国文物精华的浮雕墙体，模拟福建土楼风格的自然馆建筑，扩散水纹状的主体建筑广场。这些元素结合福建博物院三面环水的地理位置，既体现了福建的民俗风情和扬帆远行的发展愿景，也体现了水意象在中国文化中的和谐寓意；在学科层面来讲，体现了各学科融合撑起的博物馆事业发展的寓意。

For nearly 70 years, from Fujian Provincial Science Museum to Fujian People's Science Hall, to Fujian Museum, and finally to the establishment of its new Museum, Fujian Museum has been improving service environment facilities according to social development' s demand for culture, and changing service philosophy in combination of the development chance for Fujian province' s economy, and devoted to turning itself into an "open museum" providing all-round cultural services.

The main building displays Fujian style with harmonious connotation. The new Museum was completed in October, 2002, covering 6 an area of hectares, with floor area of 36,000 square meters. The main floor integrates several elements full of Fujian cultural features: " 几 " shaped roof with Fujian folk house features, totem poles embodying implications of Chinese character " 闽 ", embossment wall congregating cultural essence of China, natural exhibition hall simulating Fujian tulou style, main architectural square of spreading ripple shape. The combination of these elements and the Fujian Museum's geographic position surrounded by waters on three sides reflects Fujian local customs and its development vision for sailing far away, as well as the harmonious meaning of

01

01. 博物馆老馆旧影

对页
02. 博物馆新馆鸟瞰
03 总平面

01. Old photograph of the old hall

Opposite page
02. Aerial view of the new hall
03. Master plan

02

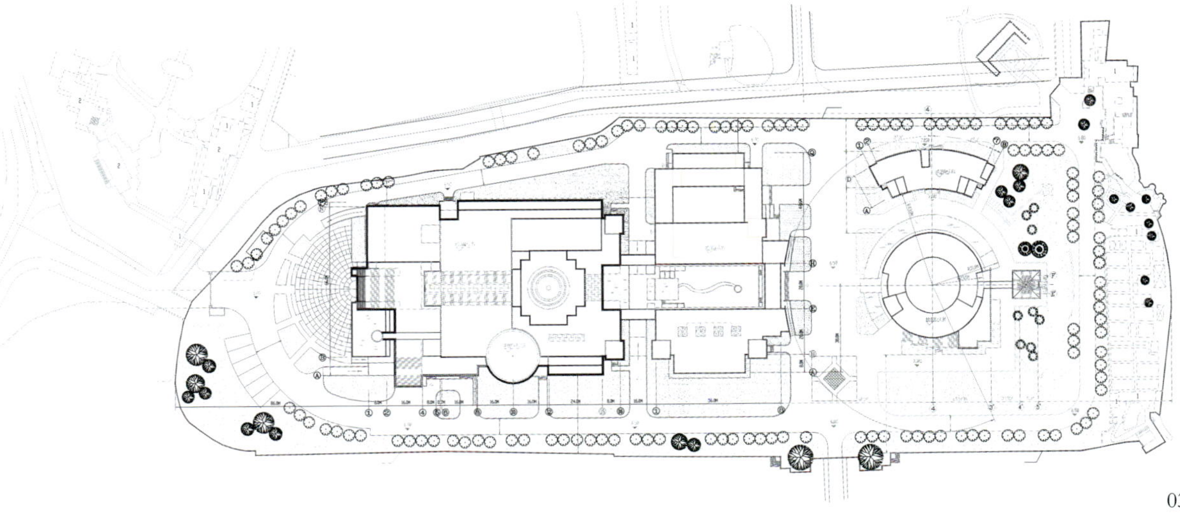

03

04

建筑采用惠安粉红麻石材，其质地和建筑外形设计贴合完美，共同烘托出博物院的典雅氛围。除了设计理念上的创新，新馆还加强了功能的拓展，集展览、教育、科研功能于一体。新馆展览面积 15000m²，共有15 个展厅。此外新馆还设有贵宾厅、会议室、具备六声道同声传译且可放映投射电视及可容纳 345 人的多功能学术报告厅和影像厅。

water imagery in Chinese culture; from the view of subject level, it reflects the importance of museum development lifted up by integration of various subjects.

The building employs Hui' an pink porrino stone, whose texture perfectly fits the building's appearance design, and highlights the museum's elegant atmosphere together. Except the innovation on design philosophy, the new exhibition hall has also strengthened the function expansion, integrating exhibition, education and scientific research into a whole. The new exhibition hall covers an area of 15,000 square meters, boasting 15 exhibition rooms. Moreover, the new exhibition is provided with VIP rooms and meeting rooms. It is equipped with 6-channel simultaneous interpretation and can broadcast projection TV. Its multi-functional academic report room and video room can accommodate 345 people.

04. 新馆正门

对页
05. 新馆侧影
06. 福建博物院新馆角楼
07. 新馆自然馆

04. Main entrance of the new hall

Opposite page
05. Side view of the new hall
06. Turret of the new hall
07. The Nature Hall of the new museum

05

06

07

China Museum for Fujian-Taiwan Kinship

中国闽台缘博物馆

中国闽台缘博物馆是反映福建与台湾历史关系的专题博物馆，坐落于福建泉州清源山风景区，占地 102 805 m^2，主体建筑面积 23 332 m^2。建筑材料采用闽台地域传统建筑特色的红砖白石。外观运用“万物归宗”的思想，采用了“天圆地方”的设计理念。屋面设置四道半弧形的看台斜阶，每道斜阶有 108 级台阶，可以直通馆顶的观景天台，作为“地方”到“天圆”的过渡，展示建筑美学的曲线美，也象征着中国闽台缘博物馆作为海峡两岸沟通交流的桥梁和纽带的寓意。屋顶的圆环形观景天台与屋面半弧斜阶之间采用了不锈钢网架结构，中间竖立的镀金和平莲花，永远传播吉祥和平的福音。

景观广场由馆名卧碑、倒影池、音乐喷泉、九龙柱、七彩柱等组成，以闽台建筑特有的红色为基调，利用天然石板材和传统地砖红白相间地分隔出醒目的图案，宛如一张巨大的地毯，与主体建筑融为一体。九龙柱和磐石卧碑构成寓意深远的组合，象征海峡两岸同胞都是龙的传人，两岸关系坚如磐石。

China Museum for Fujian-Taiwan Kinship is a special museum on the historical relationship between Fujian and Taiwan. It is located in Qingyuan scenic spot, Quanzhou, covers a site area of 102,805 square meters and the main building's floor area is 23,332 square meters; the construction material uses red brick and white stone applied in the traditional architecture of Fujian-Taiwan. The appearance utilizes the thinking that all items are from the same thing, and the design idea adopts the Sky-round and the Earth-square. The four half and a circular access ramps are designed in roof face, and each access ramp has 108 steps to the viewing platform on top of building; from the Earth-square to the Sky-round it shows the curvaceous architecture, symbolizes China Museum for Fujian-Taiwan Kinship is used as bridge and tache for communication across the Taiwan Strait. Between the top round viewing platform and access ramps is used the stainless steel reticulation structure, and in the middle stands gilded peace lotus flower, which spreading peace and luck for ever.

The square is composed of stone of museum name, reflecting pool, music conduit, nine-dragon pillar and color pillar, the basic color is red which is typical architectural color in the South of Fujian, with natural stone and traditional tile, the striking patterns have been separated by red-white colors, as if the square was a huge carpet, and it has been inosculated with the main building. The nine-dragon pillar and huge rock are combined to indicate a far-reaching significance,

01

02

03

03. 博物馆主体建筑外景
04. 轴立面图
05. 总平面图

对页
01. 九龙柱
02. 七彩柱

03. Exterior of the main building
04. Axial elevation
05. Master plan

Opposite page
01. Principal structure
02. Color pillar

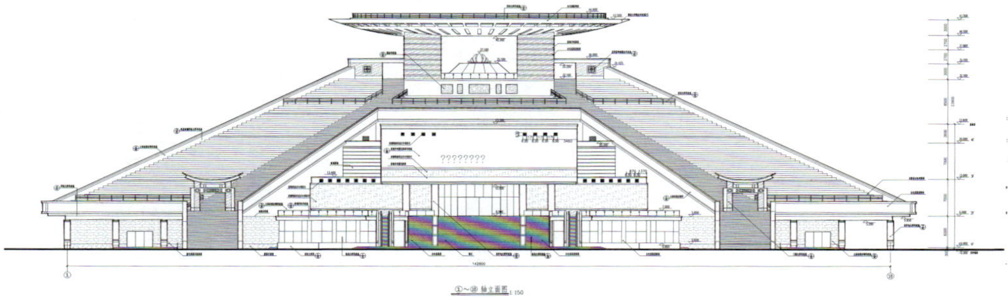

05

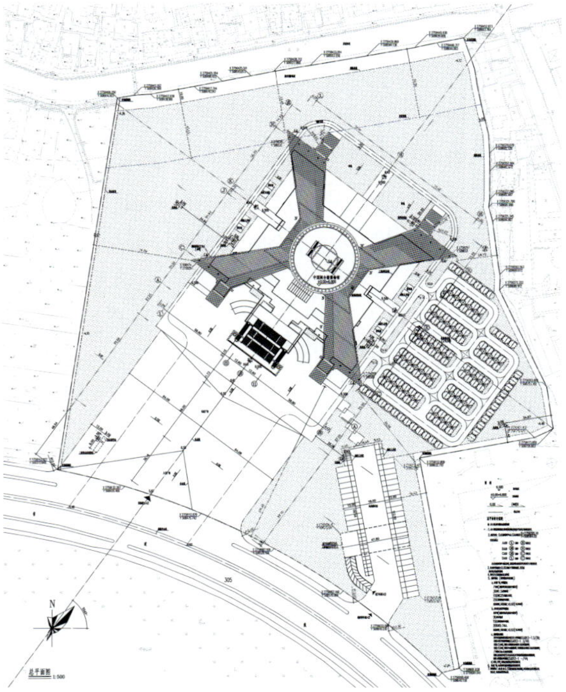

04

06

06. 大厅
07. 大厅天顶

对页
08.《闽台缘》主题展序厅
09.《乡土闽台》专题展序厅
10. 中庭回廊

06.Hall
07.The roof of hall

Opposite page
08.Preface hall
09.The preface hall of "Fujian-Taiwan Homeland"
10.Atrium cloister

博物馆二楼大门两侧墙壁各设青石浮雕，长 16 m、宽 4 m，集中展现海峡两岸民众建设共同家园的历史记忆和同文同根的民间信仰、先民图腾。

博物馆的中庭、走廊及基本陈列厅的建筑元素均运用闽台传统建筑“出砖入石”的特色工艺。同时，采用闽南特色的磨漆画、攀天龙柱、黄色琉璃顶、红色宫墙、青石墙基，使整个陈列厅既具闽台地域特色，又不失现代时尚之大气。

which symbolizes people both in mainland and Taiwan are dragon's offspring, and the relationship is as solid as rock.

On both the walls across the main gate on the second floor is carved black marble relief, 16 meters long and 4 meters wide, demonstrating the historic memory of people across the straits for building homestead and the identical folk religion and ancients? totem of the same clan.

The architecture elements of the atrium, hallway and basic show room utilize the traditional architecture technique "embodied and featured in every brick and stone". Meanwhile, the whole show room is equipped both with local feature of the South of Fujian and modern magnificence by utilizing the South-Fujian-featured lacquer picture, sky rising dragon column, yellow glazed tile roof, red palace wall and black marble wall footing.

07

08

09

10

Jiangxi Museum

江西省博物馆

江西省博物馆成立于 1953 年 3 月，原址在南昌八一广场之南，与八一起义纪念碑相互衬托。原建筑属于典型的 20 世纪 50 年代的建筑风格，并在当时属于南昌市的十大地标性建筑。半个多世纪以来，历经几代博物馆人不懈努力，江西省博物馆已成为蜚声海外的大型综合性博物馆。

为满足人们对文化精神及对历史的探索。1999 年 10 月 1 日，江西省自新中国成立以来投资最大的一项文化基础设施——江西省博物馆新馆落成并正式对外开放，这标志着江西省博物馆以崭新的姿态迈向了新世纪。

新馆坐落在南昌市赣江与抚河环抱的新洲上，东临抚河，南接中山桥，西面赣江，北依江南三大名楼之首的滕王阁，环境优美。项目占地 40 000 m^2，建筑面积 35 000 m^2，展厅面积 13 000 m^2，将历史、自然、革命三个博物馆合在一起，组成一座综合性博物馆。

Jiangxi Museum was founded in March, 1953, the former site was located to the south of Bayi Square, contrasting to the monument of the August 1 uprising. The former museum showed the typical style of the buildings of 1950s, which was known as one of the ten iconic landmarks in Nanchang City. Now Jiangxi Museum has become a large-sized comprehensive museum and enjoyed a high reputation from abroad owing to the unremitting efforts of our colleagues from generation to genertaion.

In order to satisfying people's exploration to culture and history, this new site was built-the largest invested infrastrcture in Jiangxi since the establishment of the nation, and officially opened to the public on 1st october, 1999. From then on Jiangxi Museum approached the dawning of the new century with an entirely new outlook.

Located on Xinzhou which is surrounded by the Gan River and Fu River, the new site faces the Fu River on the east, Zhongshan Bridge on the south, Gan River on the west and Prince Teng's Pavilion on the north. The museum takes up the area of 40,000 sqm with the building area of 35,000 sqm. The exhibition hall occupies 13,000 sqm, including the hall of history, hall of nature as well as hall of revolution.

01

01. 博物馆老馆旧影

01. Historical photo of the old museum

02

03

04

02-03. 外景　02-03. Exterior
04. 博物馆全貌　04. Panoroma of the museum

05

建筑主体造型取材于明代景德镇御窑出土的白瓷三管器，将历代陶瓷艺术与现代建筑风格巧妙地融合于一体，庄重典雅，别具一格。馆内中心广场矗立大型青铜雕塑“时空”，由三根仿商代青铜器的扉棱和提梁，仿商代国宝级文物立鹿四足青铜甗和伏鸟双尾虎以及火炬 4 个部分组成，寓意灿烂文明源远流长。物华天宝，人杰地灵，星星之火可以燎原。

The main structure of Jiangxi Museum imitates the white triple tube connector unearthed in the official kiln of Jingdezhen City, which combines the three halls(hall of history, hall of nature and hall of revolution) together and smartly integrates the successive ceramic culture with modern architectural style, which makes it graceful, solemn, unique and original. The huge bronze sculpture "Space-time" which erects on the central square consists of four parts: the three edged handles which are after the style of the bronze in the Shang Dynasty, the Bronze Yan(a first class cultural relics that is used to cook food in the Shang Dynasty), two-tailed tiger with a bird on the back and a torch, implying the long standing and profoundness of the splendid culture, abundant resources and brilliant people on this grand land as well as a single spark can start a prairie fire

05. 博物馆新馆全景

对页
06. 总平面

05. Panoroma of the museum

Opposite page
06. Master plan

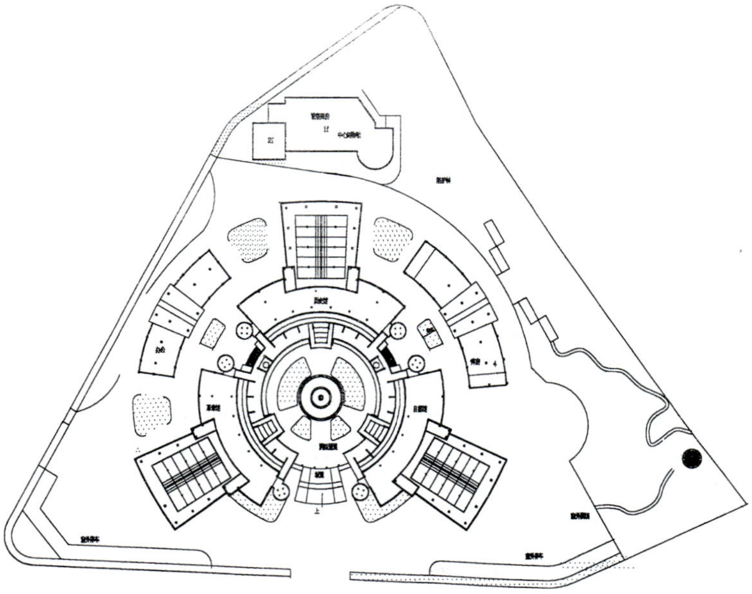

总图
PO/11
99.10.30

The Memorial Museum of the Nanchang Bayi Uprising

八一南昌起义纪念馆

八一南昌起义纪念馆位于南昌市中山路380号，所属五座革命旧址均为全国重点文物保护单位，是中宣部命名的全国首批爱国主义教育示范基地。

1927年8月1日凌晨，在以周恩来为书记的中共前敌委员会及贺龙、叶挺、朱德、刘伯承等领导下，中国共产党指导和影响的国民革命军2.3万余人在南昌举行起义，打响了武装反抗国民党反动派的第一枪，在全党和全国人民面前树立起一面革命武装斗争的旗帜，标志着中国共产党独立领导革命战争、武装夺取政权的开始和人民军队的诞生。

为了纪念南昌起义这段英雄历史，传承和弘扬八一精神，1956年，党和人民政府在南昌起义总指挥部旧址建立了八一南昌起义纪念馆，陈毅元帅题写了馆名。1997年，江泽民同志又亲笔题词："军旗升起的地方。"建军80周年之际，为加强对八一起义革命旧址的保护，更好地发挥八一南昌起义纪念馆作为爱国主义和革命传统教育基地的作用，中共南昌市委、南昌市

The Memorial Museum of the Nanchang Bayi Uprising which is located on No.380 of Zhongshan Road in Nanchang is one of five former sites for the revolution. All of five sites are key cultural relics under state protection and the first national patriotism education bases named by the Central Propaganda Department.

Early in the morning on August 1st, 1927, the revolt was held by the national revolutionary army with 23 thousand men guided and impacted by the CPC(Communist Party of China) in Nanchang, which was under the leadership of the CPC Front Committee with Zhou Enlai as clerk and He Long, Ye Ting, Zhu De, Liu Bocheng etc. The Nanchang uprising fired first for armed reacting KMT(Kuo Min Tang) and set a banner of the revolutionary armedly struggle in the presence of the CPC and the whole country, which indicated the beginning of leading the revolutionary war independently and armed forces capturing the political power by the CPC, and the birth of people's army.

In 1956, to transmit and carry forward the Bayi spirit, the Memorial Museum of the Nanchang Bayi Uprising was built at the former site of Nanchang uprising's general headquarter in commemoration of this heroic history. Marshal Chen Yi autographed the name of museum. Jiang Zemin wrote an inscription: "The military flag had been raised." in 1997. At the moment of eighty anniversary of army building, to strengthen protection for former site and play a role of the education

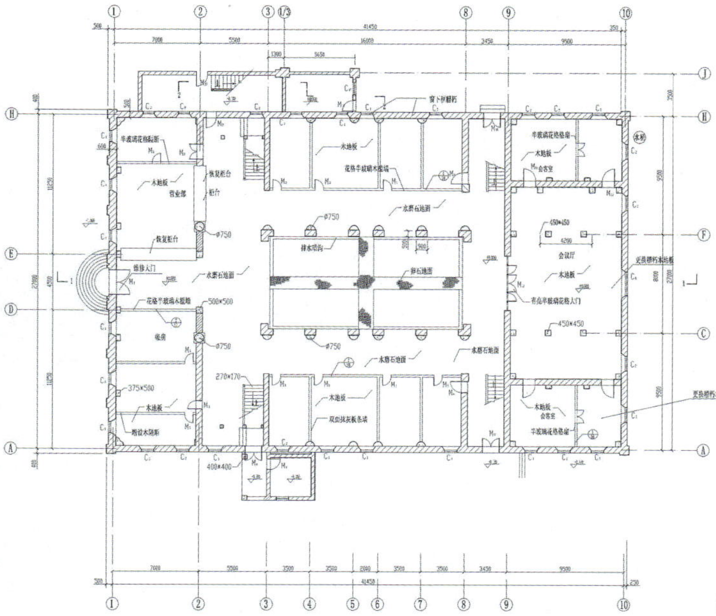

01

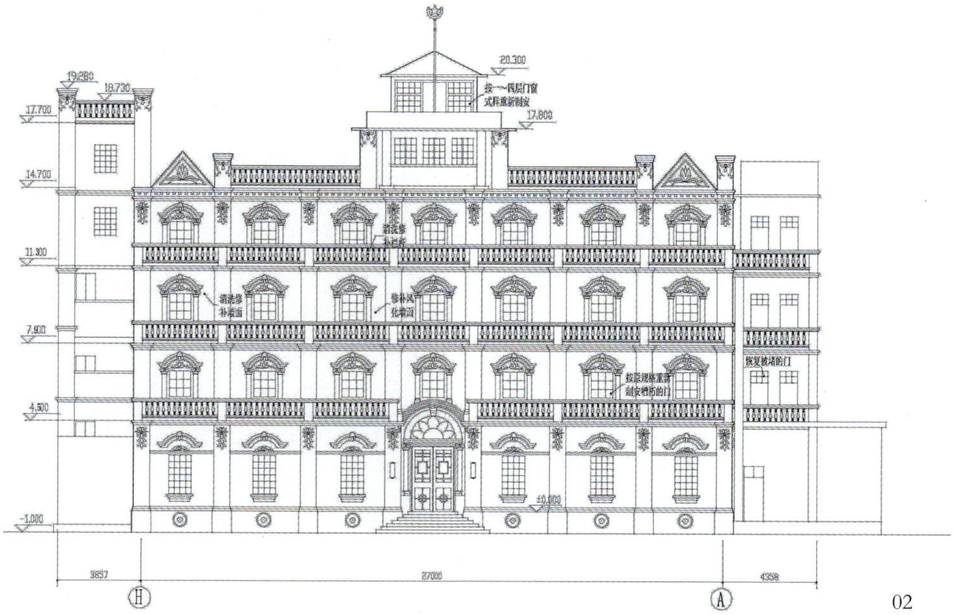

02

01. 首层平面维修图
02. 正立面维修图
03. 南昌八一起义纪念馆全景

01. Maintenance of ground floor
02. Maintenance of front Facade
03. Panorama

03

政府决定对八一南昌起义纪念馆进行改扩建，新建一座陈列大楼，采用各种先进的陈列手段全面生动地展示南昌起义这段英雄历史和人民军队发展壮大的光辉历程。同时，对八一南昌起义旧址进行维修复原，再现了当年八一起义时的历史氛围。

04

base of patriotism and revolutionary tradition, CPC Nanchang Municipal Committee and Nanchang city government decided to reconstruct and extent the memorial museum for exhibit. In the new building, the heroic history and the glorious course of people's army growing up were displayed completely and actively through all kinds of advanced measures. Meanwhile, the historical atmosphere of Bayi uprising in those days reappeared after renovation and restoration for the former sites of Nanchang Bayi Uprising.

05

06

07

04. 南昌八一起义纪念馆园区
05. 序厅：石破天惊第一枪
06. 展厅二
07. 展厅四

对页
08. 人物浮雕墙：他们从南昌起义走来
09. 喜庆礼堂
10. 周恩来工作过的 25 号房间
11. 大型壁饰型景观："上井冈之路"

04. Site of the Memorial Museum
05. Prelude Hall: The First Fire Breaking the Dark
06. Exhibition Hall 2
07. Exhibition Hall 4

Opposite page
08. Portrait Relief
09. Ceremony Hall
10. Room No. 25, Here Zhou Enlai Worked
11. Wall Landscaping: *Road to Jinggang Mountain*

08

09

10

11

The New Shandong Museum

山东新博物馆

山东新博物馆选址在济南市东部中心区域，2007 年 12 月 29 日奠基，2008 年 4 月全面开工，建设用地 140 000 m^2，主体建筑面积 83 000 m^2，人防及地下配套工程 45 000 m^2，广场约 100 000 m^2。

新馆主体建筑汲取中华传统文化“天圆地方、天人合一”的理念，气势宏伟、巍然屹立。建筑造型采用雕塑化手法，在立方体主体基础上，以四角钢结构体现出四角模数化切削的建筑风格。外墙采用超大尺度的山东泰沂山脉所独有的雪花青石材干挂，赋予建筑坚如磐石、稳如泰山的造型特点，传达出山东厚重的文化积淀和坚韧不拔的精神特质。泉水是济南的象征，以趵突喷涌为形象代表的穹顶结构，寓意着山东经济的腾飞，文化的大发展、大繁荣。

博物馆主体建筑划分为陈列展示区、藏品库区、业务科研区、综合服务区、安全保卫区、行政办公区、广场及配套设施七大部分。馆内大尺度的空间给人浑然天成的气魄，高大雄伟的大台阶展厅，把整个空

The new Shandong Museum is located in the downtown of eastern area of Jinan city. The foundation was laid in December 29th, 2007. The construction began in April, 2008. It covers a total floor area of 140,000 square meters. The main building area is 83,000 square meters in size, and the supporting facilities cover 45,000 square meters. The square covers about 100,000 square meters.

The design of the main building draws idea from traditional Chinese culture which believes in “round sky and square earth” and “harmony between heaven and man”. The design takes the sculpture method, and makes modularization cutting of the four corners of a cube. The building also shares the solid character with Mountain Tai by using huge-scale snowflake blue stones from Shandong TaiYi Mountains on its exterior walls. Its ceiling references from the Baotu Fountain, a representation of tremendous development and great prosperity of Shandong’s culture.

The main building comprises of seven sectors: the exhibition area, collection storeroom area, scientific research area, comprehensive service area, security area, administrative office area as well as the square and supporting facilities. The inner space of the building is like nature itself. The grand hall of majestic stairs extend the space to the design motif of earth and sky. The center of the ceiling is a shinning jade disk, which is the main decoration of the grand hall, as well as the representation of the square sky.

The New Shandong Museum uses a variety of new materials, technologies, and energies, such as the solar power generation system and the solid-state lighting

对页
01. 博物馆礼仪大厅
02. 新馆夜景

Opposite page
01. Ceremony Hall
02. New museum in Night

间引申为天地的设计主题。大厅中央的顶部装饰晶莹剔透的玉璧，寓意为天之圆，有浓厚的主题意味，形成大厅的中心装饰。

山东新博物馆十分注重新材料、新技术、新能源的广泛应用，新馆建筑中太阳能光伏发电技术、半导体照明技术等一系列成熟、先进的科技成果得以示范和应用，是科技含量高，优质、绿色、环保的大型公益文化建筑的典型代表，成为山东省文化、文物事业的地标性建筑。

technology. It is a typical representation of the non-profit cultural institution which has become a technological and environmental friendly landmark construction of Shandong Province's culture and cultural relics.

01

02

03. 佛教造像艺术展　03. Exhibition of Buddhist Statue Arts
04. 明代鲁王展局部　04. Exhibition of duke lu of ming dynasty
05. 汉代画像石艺术展局部　05. Exhibition of Stone Statue Art of Han Dynasty
06. 山东历史文物展厅局部　06. Exhibition of the historical culture heritages in Shangdong Province

对页　**Opposite page**
07. 山东造像艺术展　07. Exhibition of statue arts of Shandong Province

Qingdao Museum

青岛市博物馆

青岛市博物馆新馆为青岛市地标性建筑，2000 年正式对社会开放。

新馆地处崂山区政治经济文化中心，南面是青岛大剧院，东、北侧与青岛国际会展中心毗邻，西侧为崂山区文化商业中心大道。

新馆占地 68 700 m^2，主体建筑面积 26 000 m^2，三面呈外方内圆的结构，中心为圆形广场，有 74 根罗马立柱支撑。敞开处两面墙壁遥对，中心各有题为“青春”、“人生”的浅浮雕 100 m^2。主体建筑屋顶采用蓝灰色“凸”形钢瓦，层层向上向外拓展，犹如海边的浪花翻卷，体现出青岛的海洋文化元素和浪漫主义色彩。

室内建筑分为地下一层，地上三层，通高 23 m。地下一层为文物库房。地上三层呈

The new Qingdao Museum is a landmark building in Qingdao. It was formally opened to public in 2000.

Situated in the political, economic and cultural center of Laoshan District, the new museum borders Qingdao Grand Theater on the south, Qingdao International Convention Center on the north and east, and Laoshan Cultural & Commercial Center Avenue on the west.

The new museum covers a area of 68,700 square meters. The 26,000 square meters main building is square outside and circular inside on three sides, and is supported by 74 Roman-style columns. In the center of the building there is a round square. The walls at the opening are separated by 100 square meters bas-relief carvings entitled “Youth” and “Life” respectively. The roof of the building is covered with bluish grey convex steel tiles which were arranged as to give the effect of rolling waves to highlight the city features.

The inside of the building includes four storeys, three above ground and one underground, with a height of 23 meters. The underground

01

01. 博物馆圆广场

01. Round Square of the Museum

对页

02. 博物馆外景

03. 主体建筑平面

Opposite page

02. Exterior of Qingdao Museum

03. Main building plan

02

渐减叠式结构，空间宽敞，宏伟大气，拥有 16 个展厅和三条环廊展带共 8 000 m² 的陈列区，另外设有公共服务区、贵宾室、学术报告厅、办公区等。

新馆自开放以来，每年接待观众量在 30 万人以上。新馆以其独特靓丽的形象和丰富的文化内涵成为青岛市地标性建筑和著名的文化景点。

storey is used as storage for artifacts. The above-ground storeys feature a graded structure and large space to give an imposing presence. 16 exhibition halls and 3 winding corridors form an exhibition area of 8,000 square meters. In addition, there is a public service area, a VIP room, an auditorium and an office area.

Now the new museum receives over 300,000 visitors every year. It has become a new landmark as well as a famous cultural attraction in Qingdao with its beautiful image and rich cultural connotations.

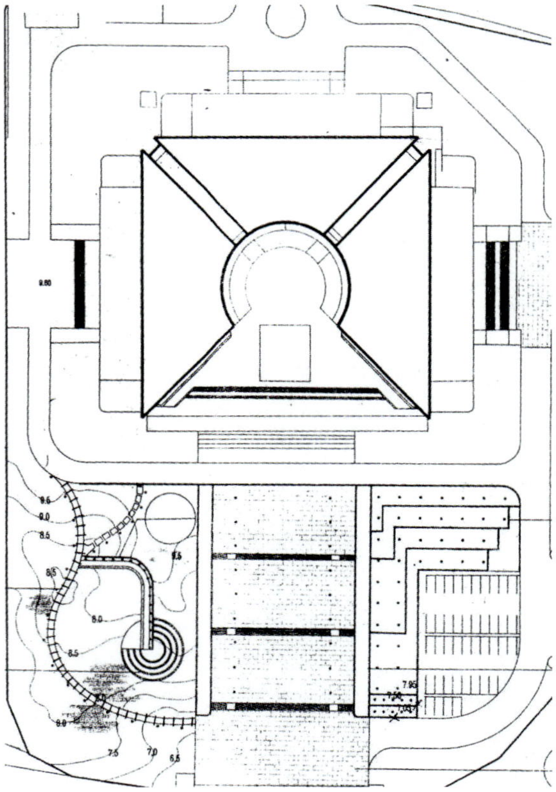

03

04

05

04. 环廊
05. 中厅

对页
06-07. 展厅

04.Winding corridor
05.Central hall

Opposite page
06-07.Exhibition hall

06

07

First Sino-Japanse War Museum

中国甲午战争博物馆

中国甲午战争博物馆位于威海市刘公岛上，是全国重点文物保护单位——刘公岛甲午战争纪念地的保护管理机构，是以北洋海军和甲午战争为中心内容的纪念遗址类博物馆。该馆建筑包括北洋海军提督署等北洋海军时期文物古建筑和新建甲午战争陈列馆两部分。

北洋海军提督署建成于 1889 年，占地面积 17 000 m^2，属典型的清代衙署古建筑群，是全国唯一保存完好的海军衙门。提督署坐北朝南，临海修筑，高踞危岩，整体按中轴线布局，前中后三进，左中右三跨，灰瓦朱柱，画栋雕梁，廊庑相接，曲曼回环。

甲午战争陈列馆占地面积逾 11000 m^2，建筑面积 7622 m^2，其中展厅面积 4500 m^2。建筑主体造型以抽象的军舰元素和具象的雕塑人物相结合，成功诠释了甲午战争的历史主题。该建筑于 2009 年获“新中国成立 60 周年百项经典暨精品工程”提名奖。

First Sino-Japanse War Museum, located on Liugong Island in Weihai city, is a key historical site under state protection as well as a protection and management institution of the Liugong Island memorial place of the war. It is a memorial site museum focusing upon Beiyang Navy and First Sino-Japanse War Museum. The museum mainly consists of two parts, the ancient structures in the time of Beiyang Navy including Beiyang Navy Headquarters and newly-built Exhibition Hall of First Sino-Japanse War Museum.

Beiyang Navy Headquarters, completed in 1889 covering an area of 17,000 square meters, is a typical ancient complex of government offices in Qing Dynasty, and the only well-preserved Navy Yamen so far. The south-facing architecture is constructed on steep coastal rocks overlooking the sea. Arranged around the central axial line, it consists of three lengthwise courtyards from front yard to back yard and three crosswise courtyards from left to right. It is exquisitely decorated with carved beams and painted rafters, green tiles contrasting with ruby pillars, and corridors and rooms are connected in zigzags and loops.

Exhibition Hall of First Sino-Japanse War Museum covers an area of 11,000 square meters and a floor space of 7,622 square meters including the 4,500 square meters exhibition area. The main part of the hall features an abstract model of a warship and a concrete sculpture figure, interpreting perfectly the historical theme of Sino-Japanese War of 1894-5. It won the Nomination Prize of "100 Classical and High-

01

01. 入口

对页

02. 陈列馆西立面

03. 北洋海军提督署正门

01. Entrance

Opposite page

02.West elevation of the exhibition hall

03.Front gate of Beiyang Navy Prefects Burea

该馆通过古建筑原状展示和新建馆综合展示，将历史遗址与文物史料有机结合起来，以独具特色的历史内涵、完善的服务设施、舒适的游览环境，得到海内外观众的普遍赞誉。

quality Projects for the 60th Anniversary of the Founding of New China" in 2009.

Museum of First Sino-Japanse War Museum organically combines historical site, relics and data by showing the original state of the ancient buildings and displaying comprehensively the newly-built hall. It has won praises from visitors home and abroad for its unique historical content, perfect services and facilities in addition to comfortable tourism environment.

Henan Museum

河南博物院

河南博物院占地面积逾 100 000 m^2，建筑面积 78 000 m^2，整个建筑群采用了传统建筑中轴对称、主从有序的布局手法。从总平面上看，在主体建筑四隅分布着八座附属建筑，聚散有致，和谐统一，寓意“九鼎定中原”。建筑群呈四面辐射状，寓意中原文明向四周传播。鸟瞰整个建筑群，宛若“大鹏展翅”向天高飞，主体建筑如大鹏身躯，展馆大门就像大鹏的头部，两侧配楼恰似大鹏的翅膀，后边与文物库房相连的过廊更像是大鹏之尾。建筑整体象征着古老中原将再现历史的辉煌，经济、文化全面振兴和腾飞。

馆区中心部位为主展馆，地面建筑高达 45.5 m，气势恢弘，上置仰斗，下接覆斗，会聚天地之精，彰显中原之气。建筑底边长 63 m，内部为四层，冠部四周分别镶嵌“四神”图案，表示古天文学中东西南北四个方位的天象星座；从建筑构造的需要考虑，建筑斜面加上白色乳钉图案，既形似青铜器上的乳钉纹样和传统建筑板门的门钉形状，又似茫茫宇宙满天星斗的反照；主体斜面四周浅蓝色的透明窗及顶部垂直而下的透明采光带，具有“黄河之水

Henan museum covers an area of 100,000 square meters, among which the floor area is 78,000 square meters. The whole architectural complex adopts a traditional axial symmetry based on which all parts of the complex are arranged to present a strong sequentiality. The plan of the whole construction is designed to imply a Chinese connotation—putting an ancient vessel which represents throne in the center predicts an ironclad dominance—by arranging eight subordinate buildings at the four corners of the main building. At the same time, the main building is balanced by its subsidiaries and presents impressive harmony overall. The radial pattern of the complex suggests that the original Chinese civilization propagate from here to all directions. From the bird's view, the architecture is just like a roc flying up into the empyreal; the main building is the body, the front door is the head, the subordinate buildings on the two sides are the wings and the corridor connected to the storehouse is the tail of the roc. The whole complex predicts that Henan with the long history will revive its past splendor and the prosperity of the culture and economy will be expected as a consequence.

The center of the complex is the main exhibition building. The part is 45.5 meters high, in which "Dou"—a traditional architectural component suspends in the center and the same component is located reversely just under it. Such arrangement signals to converge the elite of the universe and to show the honor of Central Plains. The building's side length of the four-floor exhibition is 63 meters. The

01

01. 博物院旧影

对页
02. 博物院主展馆正面
03. 博物院主展馆东侧

01. Historical photo of the museum

Opposite page
02. Front facade of the main exhibition hall
03. East side of the main exhibition hall

02

03

天上来"的磅礴气势。主体建筑造型奇特，气势宏伟，独具现代艺术风格，充分体现了源远流长、博大精深的中原文化特征。

2000 年，河南博物院主展馆荣获河南省唯一的"中国当代建筑艺术精品"和"中华百年建筑经典金榜"荣誉。

four sides of the roof are decorated with the patterns of four deities which represent the constellations in the four directions in ancient astronomy. Based on the consideration of the structure, white nails are settled on the slope of the architecture to form a special pattern, which not only imitates the nails on bronzes and doornails on traditional doors but also constructs a vivid symbol of innumerous stars in the sky. The light-blue windows around the slope and the glass lighting belt on the roof derive the idea from the Yellow River and succeed to express the majestic spirit. In sum, the main building adopts such a special design to express its modern style and at the same time to embody the profound Central Plains' distinctive culture which is of long standing.

In 2000, the main building of Henan Museum was the only architecture which won the credits of "Contemporary architectural art of China" and "Century architecture in China".

04

05

04. 博物院主展馆
05. 博物院院区

对页
06. 主展馆大厅
07. 主展馆序厅

04. Main exhibition hall of the museum
05. Courtyard

Opposite page
06. Atrium of the main exhibition hall
07. Prelude hall of the main exhibition hall

06

07

08

09

10

11

08-15. 展陈厅　08-15. Exhibition hall

12

13

14

15

Zhengzhou Museum

郑州博物馆

郑州博物馆是一座地方综合性博物馆，成立于1957年7月1日。原址设在郑州市建设东路碧沙岗公园北伐军阵亡将士烈士祠，为1928年8月冯玉祥所建，占地面积4 330 m^2。

1997年兴建新馆，1999年12月28日建成并对外开放。新馆位于郑州市嵩山南路168号，占地面积9 867 m^2，建筑面积14 200 m^2。其中主展馆8 337 m^2，是以郑州出土的商代青铜方鼎为造型基础，配以圆形碟状屋顶，取“天圆地方，鼎立中原”之寓意，融民族文化与时代精神于一体，独具特色。

在市委、市政府的关心和支持下，以展示郑州古都历史文明为主旨的基本陈列“古都郑州”于2008年5月18日（国际博物馆日）完成并对外开放。古都郑州陈列总体设计简约大气，古朴厚重，形式设计紧扣主题内容；空间规划科学，展线设置通顺流畅；互动参与设施设计精巧，生动有趣，专业性、学术性、知识性、趣味性和观赏性有机结合。该展陈对于展示郑州作为国家级历史文化名城的文化内涵，树立“中国八大古都”之一的城市文化品牌有着重要意义。

郑州博物馆始终坚持“以人为本”的服务理念，先后被授予河南省文明博物馆、河南省爱国主义教育基地、郑州市爱国主义教育基地、河南省文明单位等荣誉称号。

Zhengzhou Museum is a local comprehensive museum, founded in July 1, 1957. The former museum, which was located in the Martyrs Shrine for the Northern Expedition Dead, in the Garden of Shagang ,on Jianshe Road in Zhengzhou, was built by General Feng Yuxiang in August 1928 ,covering an area of 4,330 square meters.

The New Museum, built in 1997, was opened to the public on December 28, 1999. Situated on the No. 168 of South Songshan Road, the new building covers an area of 9,867 square meters, and has a floor area of 14,200 square meters. The main exhibition, occupied 8,337 square meters, was designed on the base of the style of traditional rectangular Ding (a kind of ancient Chinese cooking vessel) unearthed in Zhengzhou, with a circle roof of the shape of plate. The concept of the architectural design of Zhengzhou Museum is based on the traditional Chinese philosophy that “under the hemispherical dome, we stand steadily in the middle of China”, using a combination of both traditional culture and zeitgeist with its unique characteristics.

Under the concerns of municipal committe and government, the basic exhibition-“the ancient city of Zhengzhou”—was completed and opened to the public on May 18, 2008(the International Museum Day, set by ICOM),which aims to show the historical culture of Zhengzhou as a ancient city. The exhibition, designed simplicity but dignified and liberal, classical and historical, keeps to the point of the very theme. The space was divided rationally and the visiting path was set fluently. The interaction and participation devices are designed delicate, a combination of expertness, science, infotainment, playfulness and ornamentality. As a national historical cultural city, it is of great significance to display the traditional cultural meaning of Zhengzhou and establish the urban brand as one of “the great eight ancient capitals of China”.

Adhering to the "people-oriented" service concept, Zhengzhou Museum has been awarded the Museum of Civilization in Henan Province, Patriotism education base in Henan Province, Patriotism education base in Zhengzhou City and the honorary title of Civilized Units in Henan Province.

01

02

01. 博物馆外景
02. 博物馆老馆旧影（摄于 1963 年）

01. Exterior of the museum
02. Old photograph of the museum (1963)

The new hall of Luoyang Museum

洛阳博物馆新馆

洛阳博物馆新馆位于城市中轴线上，北临风景如画的洛浦公园，南接规模宏大的隋唐城遗址植物园，环境优美，交通便利。项目占地 300 亩（200 000 m²），建筑面积 62 000 m²，总投资 3.5 亿元，建筑外形如大鼎屹立，寓意“定鼎洛邑”、“鼎立天下”，体现了洛阳十三朝古都的历史内涵和特色。

新馆建筑由主楼和附楼组成。主楼两层，建筑面积 21 000 m²，一层设置中央大厅、基本陈列厅、机动展厅、互动展厅，二层设置六个专题陈列厅；附楼三层，为文物库房、办公及学术研究场所。主楼左前侧设置象征唐代天枢的观光电梯。

设施先进、功能完善的洛阳博物馆新馆工程由上海同济大学设计，2007 年 11 月 29 开工建设，2009 年 2 月建成。2009 年 4 月成功举办了世界邮票博览会，10 月举办了中国大遗址保护高峰论坛系列展览，2010 年 4 月举办了“春韵物华”专题展览。

洛阳博物馆新馆已然成为洛阳市新的地标性建筑和展现洛阳厚重历史文化的窗口。

The new hall of Luoyang Museum stands on the axis of Luoyang, with the picturesque Luopu Park in the north and the huge arboretum on the relic of the old city in Sui and Tang Dynasty in the south, as well as the beautiful environment and the convenient transportation. The site area of the project is 300mu(200,000 square meters) and the floor area is 62,000 square meters, while the investment is 350 million yuan. The appearance of the architecture seems like a big standing ancient vessel which means"making Luoyang the capital of the dynasty" and "making the dynasty the center of the world", which shows the historical characteristics of this old capital city of thirteen dynasties.

The new hall is combined by the main building and the assistant building. The main building, whose floor area is 21,000 square meters, has two floors–the ground floor contains the central hall, basic display exhibition hall, flexible exhibition hall, interactive exhibition hall, and the second floor contains six special exhibition halls. The assistant building has three floors used as the store room of the cultural relics, the office rooms and the academic place. There is an observation elevator in the left side of the main building symbolizing the Tianshu in Tang Dynasty.

Designed by Tongji University of Shanghai, the construction in the new hall of Luoyang museum started in November 29th, 2007 and was completed in February 2009, which made the museum as a whole new place with advanced facilities and functions. The Museum successfully hosted the World Stamp Expo in April 2009, the Chinese Summit Forum and series of Exhibitions of Site Protection in October 2009, and the Special Exhibition of "Treasures with the spirit of spring" in April 2010.

The new hall of Luoyang Museum has become the new landmark of Luoyang and the window for the city to present its long historical culture.

对页
01. 新馆内景
02. 新馆外景

Opposite page
01. Interior of the new museum
02. Exterior of the new museum

01

02

03

04

03. 永恒的文明展厅
04. 宫廷文物展展厅（家具）

对页
05. 洛阳文物珍宝展
06. 中国古代都城文明展

03. Exhibition Hall of the Civilization
04. Exhibition of Royal Relics (Furniture)

Opposite page
05. Treasures Exhibition of Luoyang Relics
06. Exhibition of Civilization of Chinese Ancient Capital

05

06

National Museum of Chinese Writing

中国文字博物馆

中国文字博物馆字坊高 18.8 m，宽 10 m，取甲骨文、金文中“字”之形，是中国文字博物馆的象征性建筑之一。

字坊前方两侧各有金色铜质凤鸟雕塑一尊，取材于商周青铜器上常见的凤鸟纹纹饰，整体形象高贵典雅，有待翔之势，象征着新时期我国文字文明新飞跃。

在通向主体馆的主干道两侧是由 28 片极具代表性的铜质甲骨片组成的碑林，隐含了殷商时期最具代表性的两种元素——甲骨文和青铜器，它代表着 28 星宿，象征人与自然和谐同一，是我国古典哲学中“天人合一”思想的生动体现。

中国文字博物馆主体馆高 32.5 m，建筑设计采用“后现代”理念，造型定位取材于殷商甲骨文、金文所概括的最富有哲理、最经典、最神圣的建筑形象——象形文字的“墉”字。设计方案和造型包含了殷商时期的高极宫殿建筑形象的基本要素，采用殷商时期的饕餮纹、蟠螭纹图案浮雕金顶，展现殷商宫殿“四阿重屋”的效果，不仅隐含了中国文字的文化内涵，也显

As one of the symbolic architectures of National Museum of Chinese Writing, Zifang (a memorial archway shaped after Chinese charater “字”)is 18.8 meters high and 10 meters wide, and the shape of “字” in oracle-bone and bronze-ware inscription is where the design comes from. In front of both sides of Zifang are bronze sculptures of Chinese phoenix which get its inspiration from the decorative pattern of phoenix usually found in the bronze wares of Shang and Zhou dynasties. With their splendid and elegant appearance and be-about-to-fly posture, the sculptures symbolize the new advancement of Chinese writing civilization in a new stage.

The trunk to the central museum is lined on both sides with a group of steles composed of 28 representitive bronze oracle bones, which contain the oracle-bone inscription and bronze ware, two most typical elements in late Shang dynasty, representing the 28 constellations. Symbolizing the harmony between the human being and nature, the design vividly embodied the thought of combination of the nature and human being in Chinese classic phylosophy.

The central museum is 32.5 meters high. Adopting the idea of post-modernism in architectural design, the central museum was constructed after the pictograph “墉” . The most classical and sacred architectural form induced from the oracle-bone and bronze-ware inscription. With the basic elements used in palace of late Shang dynasty included design and construction, the central museum

01

01. 凤鸟

对页
02. 主体馆外景
03. 总平面
04. 立面

01. Phoenix sculpture

Opposite page
02. Exterior of the central museum
03. Master plan
04. Elevation

02

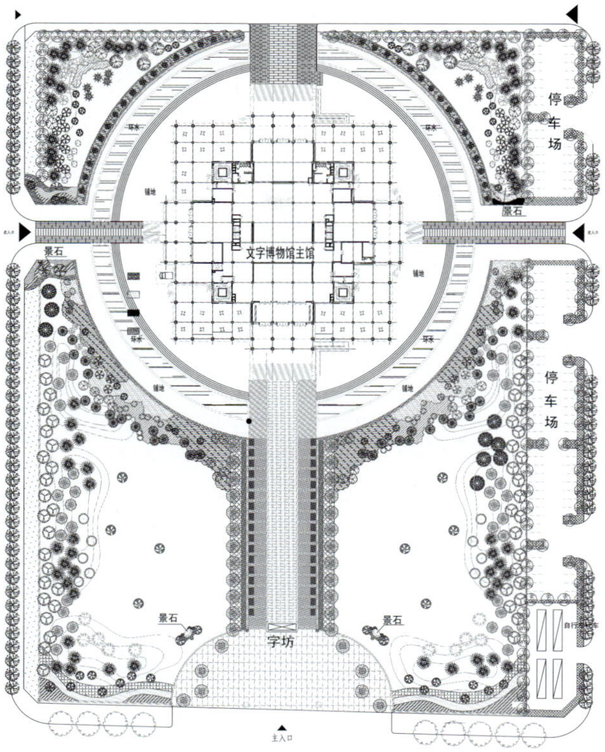

03

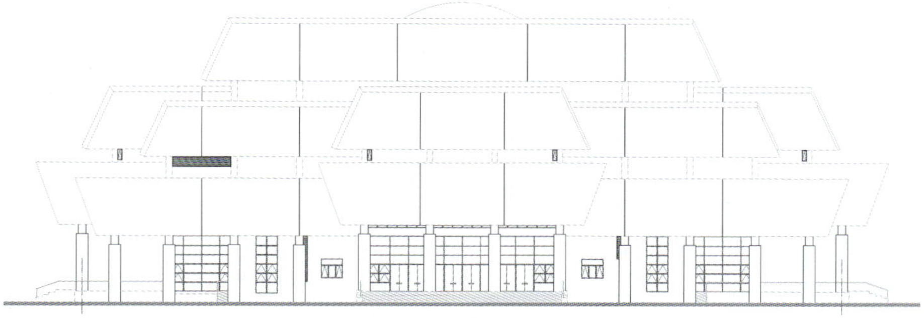

轴立面图 1:150

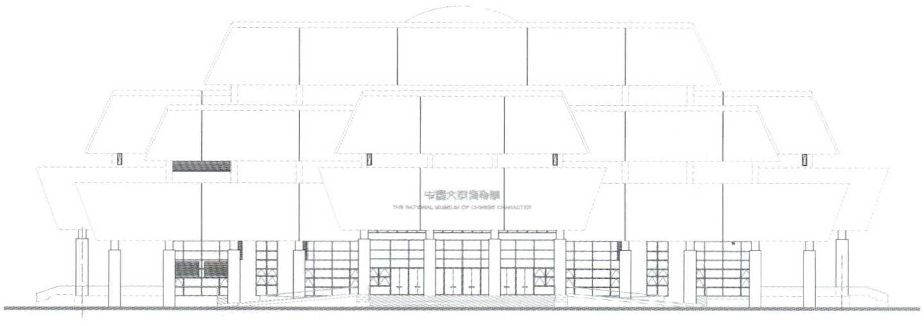

轴立面图 1:150

04

示了文字在我国文明发展长河中举足轻重的地位。中国文字博物馆主体馆把金碧辉煌的建筑外观与错落雅致的外部环境融为一体，显示了人文与自然高度和谐的哲学内涵。

adopted the Taotie and Panchi patterns in embossment of roof (Taotie and Panchi are two sons of dragon according to the legend), and displayed the characteristics of palace of late Shang dynasty, that is, the roof is constructed with four slopes and double eaves. So the centual museum not only kept the Chinese writing culture within the construction, but also represented the significant role of Chinese writing in the developmental history of Chinese civilization. Combining the splendid appearance and refined outer environment, the central museum of National Museum of Chinese Writing displayed the harmony between the civilization and the nature.

05

07

06

08

09

08. 夜景
09. 铜质甲骨片组成的碑林

对页
05-06. 博物馆浮雕金顶
07. 细部

08. Nightscape
09. Steles composed of bronze orale bones

Opposite page
05-06. Embossment of roof
07. Details

Hubei Provincial Museum

湖北省博物馆

湖北省博物馆是湖北最重要的文物收藏、研究和展示机构。它筹建于 1953 年，其 3 000 m^2 的旧馆于 1963 年建成并对外开放。1999 年 1 月，湖北省政府投资 3 000 万元兴建了 5 717 m^2 的编钟馆。2002 年 11 月，国家发改委批复了湖北省博物馆扩建工程初步设计，建成后总建筑面积 51 580 m^2，其中展厅面积 12 384 m^2，文物库房面积 6 872 m^2。

湖北省博物馆扩建工程设计为新古典风格，设计结合现代技术、现代工艺和材料，运用现代审美意识，对传统建筑形式进行再创造。建筑突出地表现了战国高台建筑的特色和楚文化大气、精美、浪漫的艺术特点。建筑外墙为浅灰色花岗石，屋面采用青灰色琉璃瓦。博物馆以其恢弘的空间构图和独特的建筑造型，成为风景秀丽的东湖之滨一颗灿烂的明珠。

主体建筑采用对称的轴线构图，由综合陈列馆和编钟馆、楚文化馆与主入口门楼共同形成“楚宫双阙对阳台”的完整格局。建筑造型结合楚式高台建筑的特点加以创

Hubei Provincial Museum is the most important institution for collection, research and exhibition of cultural relics in Hubei Province. The preparation for the construction of the museum started in 1953. The old hall of 3,000 square meters was completed and opened to the public in 1963. In January, 1999, the provincial government invested 30 million yuan for the construction of the Chime-bells Hall of 5,717 square meters. In November, 2002, the National Development and Reform Commission approved the preliminary design of the extension project of the museum. The museum covers an area of 51,580 square meters after the extension project was completed, including an exhibition hall of 12,384 square meters and a storeroom for cultural relics of 6,872 square meters.

The neoclassic design of the extension project of the museum integrates the modern technique, technology and material to make recreation based on the traditional architectural forms with a modern aesthetic consciousness. The building gives an eminent manifestation of the unique features of the high-terrace buildings in the period of Warring States and the artistic characteristics of Chu culture: splendid, elegant and romantic. The exterior wall is built with light grey granite and the roofing is made of steel grey glazed tile. With the magnificent space composition and unique architectural form, the museum becomes a brilliant pearl beside the beautiful East Lake.

01

01.2002 年总体规划

对页
02. 组合正立面
03. 组合侧立面
04. 总平面
05. 湖北省博物馆鸟瞰

01.Overall planning in 2002

Opposite page
02. Front elevation
03. Side elevation
04. Master plan
05. Aerial view of Hubei Provincial Museum

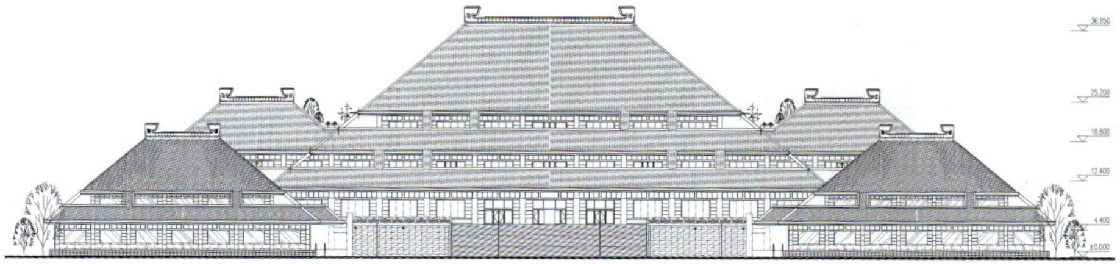

正 立 面 图

02

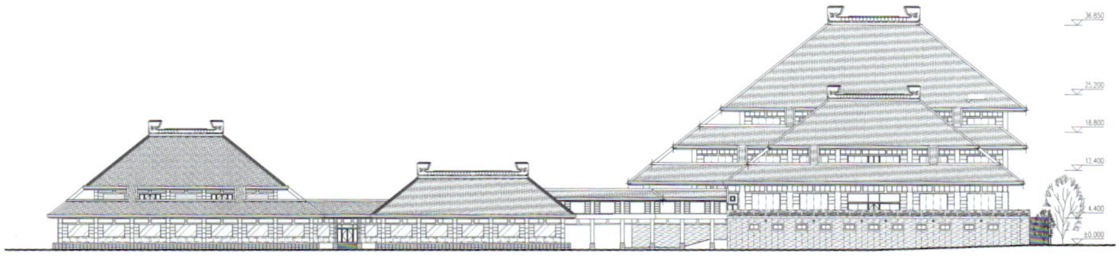

侧 立 面 图

03

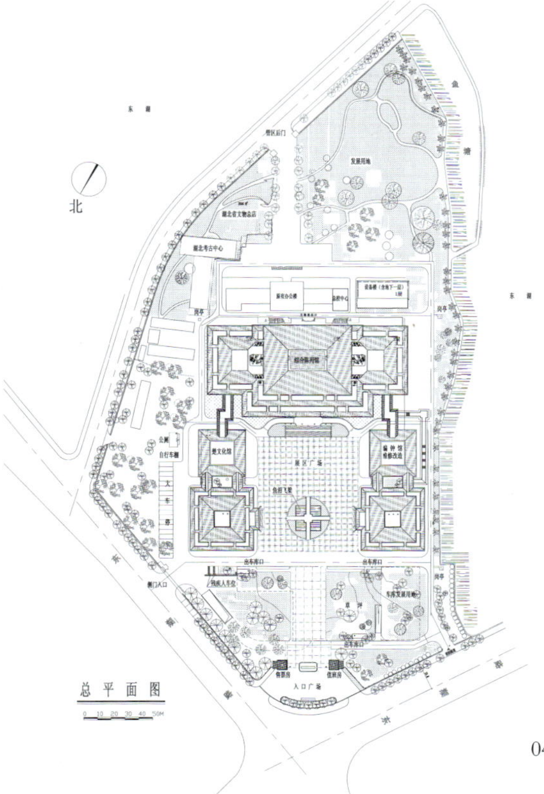

04

05

06

新。主体建筑屋面以传统的坡顶建筑为基础，在覆斗形的体量上作减法构成，结合各层的景观外廊，形成具有整体感和独特个性的建筑样式。细部设计精益求精，从建筑的基座、台阶、栏杆、墙身、门窗、檐口、脊饰到室外环境的各个方面，均精心推敲，力求达到古朴而有新意、简洁又耐人寻味的效果。

设计注重生态环境，如大面积可渗透硬质广场的设计，尽量保留地上原生景观和植被的地下车库设计，利用中庭和外墙电动百叶调节室内气候的设计，采用创造性地采用双层虑光玻璃顶棚过滤紫外线、引入自然采光的设计等，提高了建筑的可持续能力。

A symmetrical axial composition is adopted for the main building. The Comprehensive Exhibition Hall, the Chime-bells Hall, the Chu Culture Exhibition Hall and the gate tower at the main entrance form a complete pattern which was described by a poet in Tang Dynasty, "On both sides of the Chu Palace stand the watchtowers on the opposite of the balcony." It is an architectural form created in combination of the features of Chu-style high-terrace buildings. The traditional slope-roof structure is taken as the basis for the design of the roofing of the main building. The overturned hopper-shaped body where subtraction composition is made, in addition to the landscape lanais at each storey, there is an architectural from with both integrity and its unique feature. The detail design is perfect: the foundation support, steps, handrails, walls, doors, windows, cornice, cresting and all details of exterior environment are all perfectly designed to achieve the primitive yet original, simple yet thought-provoking effect.

The design attaches importance to the ecological environment. For example, a large area can be combined with permeable hard square; possible efforts are made on the preservation of the native landscape and the vegetation in the design of the underground garage; electrical

06. 主入口门阙设计

对页
07. 楚文化馆立面
08. 一层组合平面
09. 室外楼梯栏杆

06. Design of the Gate Tower at the Main Entrance

Opposite page
07.Facade of the Chu Culture Exhibition Hall
08. Composed Plan of 1s floor
09 .Outdoor Stair Railing

07

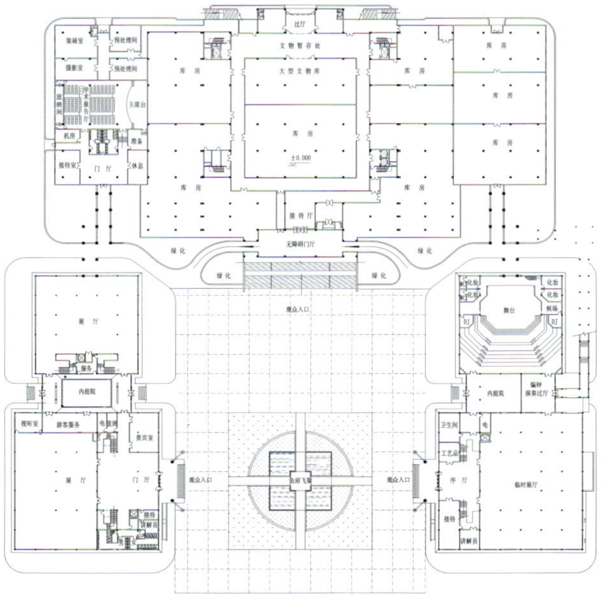

一层平面图

08

09

先进的空调、消防、安防设施的配套以及数字化模拟编钟演奏、触屏立体旋转文物展示、激光场景等先进陈列手段的运用，使博物馆的总体水平达到国家一流。

venetian blind is used at the atrium and the exterior wall to control the indoor climate; the double-glazing filter glass ceiling is creatively adopted to filter the ultraviolet rays; the design of natural lighting is adopted; etc.. All these designs increase the sustainability of the building.

The advanced air-conditioning, fire protection and safety protection facilities, in addition to the application of such advanced display technologies as the digital simulation of chime performance, the touch-screen for three-dimensional rotary display of cultural relics, the laser scene, etc., make it a first-class museum all over the country.

10. 入口区环境鸟瞰
11. 内部广场

对页
12. 中庭

10.Aerial view of the environment in Entrance Area
11.Indoor Square

Opposite page
12. Atrium

12

Jingzhou Museum

荆州博物馆

荆州博物馆位于国家历史文化名城——荆州城内，占地面积 5 万 m²，其中建筑面积 3 万 m²，绿化面积 11 000m²。建筑布局是以古建筑开元观为依托，向东、北两面扩展，现有陈列楼、珍品馆、临时展区、考古楼、办公楼、文物库房等；建筑风格为现代建筑材料仿明清时期的官式或园林式建筑，大部分建筑的檐下额枋上饰以楚国丝织品、漆木器等出土文物彩绘花纹，具有浓郁的荆楚文化特色。

开元观为全国重点文物保护单位，始建于唐开元年间，历代间有修葺，现存建筑主要为明、清两代所建。中轴线上保存有山门、雷祖殿、三清殿、祖师殿。 山门为木构门楼，雷祖殿为清式硬山顶，三清殿为单檐歇山顶，祖师殿建于高台之上，为重檐歇山顶建筑。

陈列楼位于博物馆园区中央，是一座仿明清时期宫廷建筑风格的高台殿堂式建筑。南面设双层月台和前后门廊，主体为三层单檐歇山顶主楼，两翼为两层盝顶辅楼，皆覆盖绿色琉璃瓦。陈列楼内设 5 个展厅，

Jingzhou Museum is located in the west of the famous ancient city—Jingzhou. In the southwest corner of the museum there is an ancient Taoist temple named Kaiyuan Temple. Besides, there is the Display Building, Treasure House, Temporary Exhibition Hall, the Archaeology Building, the Office Building and so on. All the architecture in Jingzhou Museum is in ancient style, and the patterns come from the lacquer and the silk which is excavated in the Chu Tombs in Jingzhou. The whole museum covers an area of more than 50,000 m².

Kaiyuan taoist temple is a national cultural heritage. It was built between Tang Kaiyuan years. It has been repaired sometimes in the past dynasties before. Those existing buildings now are built mainly during the Ming or Qing Dynasties. On the central axis there are Gate, Leizu Hall, San Qingdian, and Patriarch Palace. Gate is made by wood. Lei Zu Hal has Hard Hill roof of Qing style. San Qingdian has Xie Shan roof with Single eave. Patriarch Palace which has Xie Shan roof with double eaves is built on the top of high pedestal.

The display building is located in the center of the yard, and faces the north. It's a palace like architecture in Ming and Qing style with three floors. There are all together five exhibition halls in it, such as "The Exhibition of Relics in Prehistory","The Exhibition of Bronze in Jianghan Plain","The Exhibition of Bamboo Slips Excavated in Jingzhou","The Exhibition of Jade Excavated in Xiongjiazhong Tomb" and "The Treasure of china in Jingzhou Museum".

01

01. 博物馆旧影（摄于 1984 年）

对页
02. 总平面
03. 陈列大楼

01. Old photograph of the museum (1984)

Opposite page
02. Master plan
03. Exhibition building

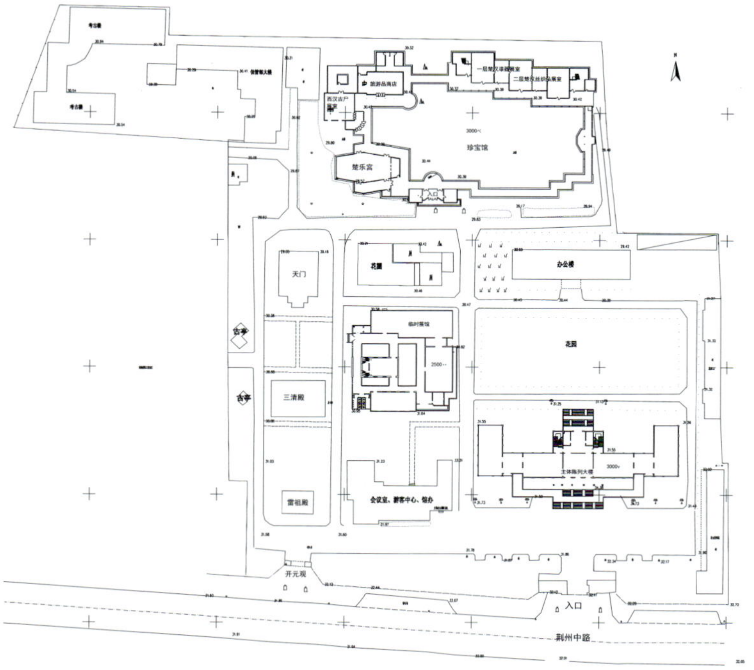
旅游品商店
一层楚汉漆器展室
二层楚汉丝织品展室
西汉古尸展室
珍宝馆
楚乐宫
入口
天门
花圃
办公楼
古亭
临时展馆
花园
三清殿
古亭
主体陈列大楼
雷祖殿
会议室、接待中心、馆办
开元观
入口
荆州中路

02

03

现有“江汉平原原始文化展”、“江汉平原青铜文化展”、“荆州楚墓暨熊家冢出土玉器展”、“荆州出土简牍文字展”、“荆州博物馆馆藏瓷器精品展”5 个常设展览。

珍品馆位于馆区北部，是一座水上仿古园林式建筑，由门厅、水池、曲桥、水榭、亭阁、廊房等组成，布局精巧，错落有致。珍品馆内举办了“凤凰山 168 号汉墓展”、“古代漆木器精品展”、“楚汉织绣品展”3 个常设展览。

临时展区是一座四合院式有回廊、天井的仿古园林式建筑，内设 3 个临时展厅，用于举办引进的外展、考古新发现展、历代艺术品展等临时展览。

荆州博物馆园区内现有碧柏、拱桐、紫薇、红叶石楠、桂花、银杏、五针松等 10 余种名贵树种和各类花卉，一年四季草木葱翠，鲜花盛开。漫步在红墙绿瓦下，徜徉于红花绿叶中，穿梭在时空走廊间，令人心旷神怡、留连忘返。

The Treasure House is located in the north of the museum. It's an over-water garden architecture with pavilions, towers and terrace halls. There are all together three exhibition halls in it, such as “No.168 Fenghuangshan Tomb Exhibition—the wet corpse in Han Dynasty”, “The Treasure of Lacquer” and “The Silk Fabrics in Warring States Period and Han Dynasty”.

The temporary exhibition hall is a Chinese traditional rectangular courtyard with ambulatory. There are three exhibition halls in it.

The whole museum forms a group of architecture in Ming and Qing style, and the comfortable environment attracts lots of visitors both at home and abroad.

04

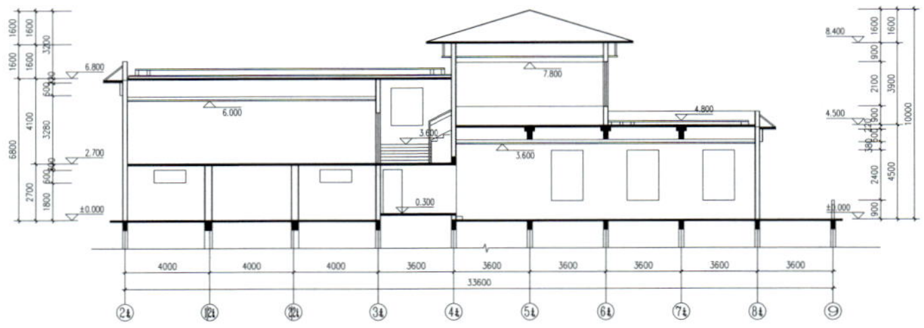

05

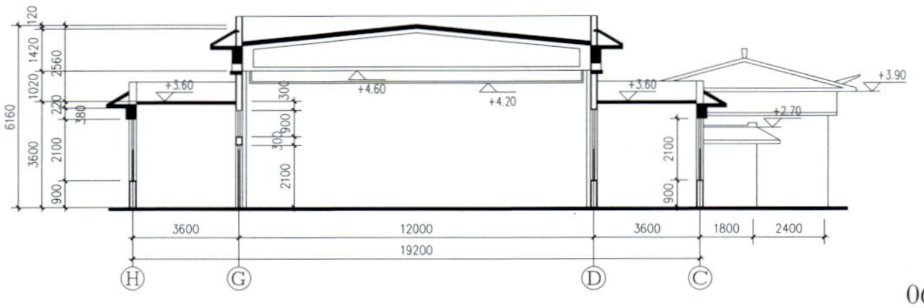

06

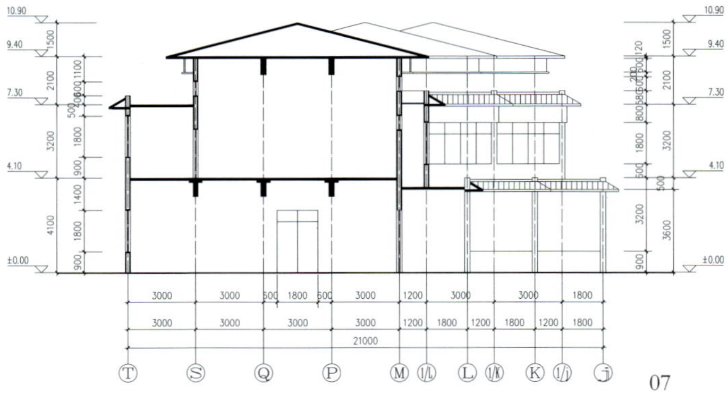

07

04. 珍品馆
05-07. 剖面

04.Treasure Hall
05-07.Section

对页
08-09. 展陈厅

Opposite page
08-09.Exhibition Hall

08

前言
江汉平原包涵了长江中游与汉水流域的广袤地区，作为古代中国的区域文化中心，一直是物阜民丰、九州通衢、兵家必争之地。
发达的江汉平原原始文化在夏、商、西周时期逐步融合，发展成为具有鲜明特色的地域文化，经过春秋争霸和战国征伐，楚国成为“地方五千里，带甲百万”的第一大诸侯国，创造了高度发达的楚文化，对秦汉时期中华民族的大一统、大融合产生了积极而深远的影响
历经岁月的沧桑，拂去历史的尘埃，青铜器——中国古代先民勤劳与智慧的结晶，就是往昔辉煌的忠实见证者之一
本展厅展出的是江汉平原考古发掘出土古代青铜器中的精华，流连于这些国宝之间，我们感受着文明的博大，激荡着民族的雄魂，探索着前进的道路……

09

Mao Zedong Relic Museum

毛泽东遗物馆

毛泽东遗物馆坐落在新中国开国领袖毛泽东的家乡——湖南韶山。2008 年 12 月 20 日建成，总建筑面积 19 000 m²，其中陈列布展面积 8 400 m²。该馆依山就势，层层迭落，呈分散式院落组合布局，充满了诗情画意的江南园林特色，白墙灰檐，又使之与韶山乡村自然和人文环境相得益彰。

遗物馆在建筑风格上融合了中国传统建筑与现代建筑的特点，双坡屋顶、风火墙、吊脚楼，这些具有湖南民居特点的元素通过清淡素雅的外墙装饰、钢制屋梁、钢材和玻璃搭起的小坡面屋顶、方钢与实木组合的窗棂等现代建筑的材料和手法加以了演绎。该馆建筑色调为灰色系，灰蓝色的屋顶、风火墙上深灰色的槽钢压顶以及青砖贴面，营造出庄严肃穆的氛围。

遗物馆展厅由轻钢、玻璃、木格栅吊顶和菱形花窗共同构成，上方为玻璃天窗，自然采光。大空间、少廊柱的设计使展线流畅，空间分割上又呈弧形变化，法国木纹石展墙优美厚重。

Mao Zedong Relic Museum is located in Shaoshan of Hunan province, the hometown of Mao Zedong, who is China's first leader since the foundation of the People's Republic of China. Built on December 20, 2008, it covers a total area of 19,000 square meters, of which 8,400 square meters is for exhibiting. Making use of hills on one side, the museum arranges its buildings in tiers and presents the layout of scattered halls integrated as a whole, a distinguishing style of idyllic gardens in South of theYangtze River. Meanwhile its white walls and grey eaves make the museum a perfect match and complement to the natural countryside scenery and the cultural environment in Shaoshan.

The museum architecture combines both traditional and modern Chinese styles. Elements with characteristic architectural style of Hunan, such as the double-pitch roof, the fire wall, the pillar-supported dwelling, etc., open up before eyes through modern construction materials and techniques like the simple but elegant outer wall decoration, steel beams, the sloping roof structured with steel and glass, the window lattice composed of square steel and solid wood, etc.. The grey-toned museum with the bice roof, the dark grey channel steel coping on the fire wall and the blue brick facing creates a solemn and serene atmosphere.

Each hall of the museum, comprised of light steel, glass, the suspended ceiling with wood grids and the diamond-shaped lattice window, makes

01

01. 博物馆院内的竹林

对页
02. 博物馆俯视
03. 全景

01.Bamboos in the courtyard of the museum

Opposite page
02.Aerial view of the museum
03.Panorama

02

03

04

遗物馆共展示毛泽东遗物千余件，一楼为“风范长存——毛泽东遗物展”基本陈列，二楼为专题展览和临时展览区。出二楼展厅，即到缅怀广场，可将各个层面的建筑尽收眼底。缅怀广场的核心部分为一个巨大水池，正中耸立着一个边长 1.5 m 的方形青铜大印，四面均刻有“毛泽东”三个大字。在缅怀广场东侧的报告厅外庭院墙面上镌刻着毛泽东手迹《沁园春 · 雪》，成为烘托建筑恢弘气势的点睛之笔。

use of natural light through the glass skylight on the roof. The design of more space but fewer pillars facilitates a smoothly-stretched viewing route for visitors and a crescent spreading trend in space division. In addition, the hall walls, which are made of a kind of French stone with a wood grain look, are very elegant and solid.

The museum displays a total of about one thousand pieces of Mao Zedong's relics, the basic exhibition "Long Live Chairman Mao–Mao Zedong's Personal Items" lying in the first floor while theme exhibitions and the temporary exhibiting area in the second floor. The exit of the hall in the second floor leads to the Recalling Square, where architectures from all corners are in full view. The center of the square is a huge pool, amidst which emerges a 1.5 meters-square bronze seal with three Chinese characters "毛泽东" (Mao Zedong) on each of its four sides. On the wall of the yard outside the auditorium, east of the square, is an engraved script of *Snow* in Mao Zedong's handwriting, which serves as a finishing touch highlighting the magnificence of the architecture.

04-05. 水景

04-05. Water scupe

05

Guangdong Museum

广东省博物馆

广东省博物馆是华南地区最大的省级综合性博物馆，于 1957 年开始筹建，1959 年正式对外开放。原址位于广州市文明路“广东贡院”旧址，现址位于广州新城市中轴线——珠江新城中心区南部的文化艺术广场（珠江东路 2 号）。

广东省博物馆新馆规划于 2002 年，2003 年正式得到广东省人民政府批准立项，并被列为广东省重点文化建设项目和建设文化大省三大标志性文化设施之一。2004 年 12 月 12 日奠基兴建，2010 年 5 月 18 日建成并向公众免费开放。建筑总投资 8.842 亿元，总用地面积 41 000 m^2，总建筑面积 67 000 m^2。

广东省博物馆的设计构思源自中国精雕细琢的传统宝盒，蕴含着传统文化中的宝盒、象牙球、玉碗、铜鼎等多种意象，意味着盛载多年累积的历史传统、智慧及文化。

馆内的空间组织概念则取材于广东传统的工艺品——象牙球工艺技术，整个空间组织就像象牙球镂空的工艺，多层次、复杂而细致。

Guangdong Museum is the largest comprehensive museum in South China, which was constructed in 1957 and opened to the public in 1959. It was located in Wenming Road, and now in the new city axis in Guangzhou—Zhujiang Dong Lu No.2 in Zhuangjiang New Town.

The new building was planned in 2002, and was formally approved by Guangdong Provincial Government in 2003. It is one of the three most important cultural projects of Guangdong Province. It started to be built on December 12, 2004, and opened to the public for free on May 18, 2010. The new building covers an area of 41,000 square meters and the floor area is 67,000 square meters, with the total investment of 884.2 million yuan.

The design concept of Guangdong Museum originated from traditional Chinese treasures, containing the box of treasures, ivory balls, jade bowl, tripod and other images, which mean holding on years of historical tradition, wisdom and culture.

The spatial organization of the museum is derived from the concept of the traditional craft in Guangdong—ivory ball technology. Just like the hollow ivory ball, the building demonstrates a variety of multilayered, complex and delicate features.

01

01. 博物馆旧影（摄于 1959 年）

对页
02. 正面全景
03. 结构剖析
04. 幕墙细部

01.Historical photo of the museum(1959)

Opposite page
02.Panorama
03.Structure analysis
04.Details of the curtain wall

02

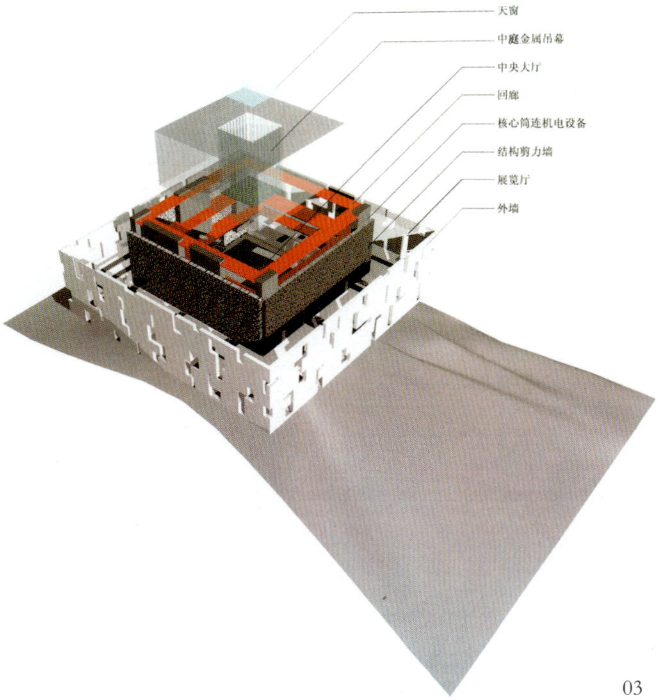

03

04

由于展馆是由巨型钢桁架悬吊出来的，中间没有一根结构柱，为陈列展示提供了很好的条件。由钢架作为主要支撑的“宝盒”四边“浮起”，离地 8 m，外表黝黑的巨型宝盒从地上腾空浮起，既具现代几何色彩味道，亦不失中国传统宝盒的精致。

该馆主要配置有展馆（20 500m²）、藏品保藏系统（8 300m²）、教育服务设施（3 100m²）、业务科研设施（4 700m²）以及安防、公共服务、综合管理系统等。展馆分为历史馆、自然馆、艺术馆和临时展馆四大部分。

Since the exhibition hall is suspended by giant steel truss, there is no column in the middle, which provides quite an advantageous condition for exhibitions. Supported by a steel frame, the four sides of “treasure box” float 8 meters off the ground, integrating both the taste of modern geometry and the fine traditional Chinese treasures.

There are the exhibition hall (20,500 square meters), collections preservation systems (8,300 square meters), educational facilities (3,100 square meters), research facilities (4,700 square meters) and security, public services and integrated management systems, etc. in the museum. The exhibition hall includes those of history, nature, art and temporary exhibitions.

05

06

05-06. 大堂

对页
07. 广东历史文化陈列区
08. 广东自然资源展览区

05-06.Hall

Opposite page
07.Guangdong History and Culture Exhibiton Hall
08.Guangdong Natral Resources Exhibition Hall

07

08

Shenzhen Museum

深圳博物馆

深圳博物馆建成于 1981 年，是深圳市建成时间最早、收藏文物数量最多、展览面积最大、接待观众量最多和设备、设施最先进齐全的博物馆，也是深圳市唯一家市级博物馆，现有新、老两处馆区，老馆为艺术自然馆，新馆为历史民俗馆。

艺术自然馆位于深圳市的中心干道深南中路，紧邻深圳市委和荔枝公园，占地 37 000 m²，建筑面积 18 000 m²，1984 年动工兴建，1988 年 11 月建成开馆。主体展楼建筑风格为传统与现代相结合，是一座 4 层的古堡式建筑。外表用米黄色泰山砖镶嵌，庄重典雅，古朴大方。大门两侧镶饰有花岗岩壁雕“古代耕战图”，门前喷水池矗立着明代十二生肖石雕，池内水清鱼跃。展楼内采用透明磨砂玻璃将自然光引入中央大厅，青色页岩石材铺地，淡雅古朴，以中央大堂为中心，螺旋式分布着 32 个展厅。

Established in 1981, Shenzhen Museum is the earliest established museum with most collected cultural relics, largest exhibition area, most visitors and most advanced facilities in Shenzhen. It is the only state-owned municipal museum of the city with two parts. One Shenzhen Museum of Art and Natural History, the other is Shenzhen Museum of History and Folk Culture.

Shenzhen Museum of Art and Natural History is located on Shennan Boulevard, bordering Lychee Park and CPC Shenzhen Municipal Committee. The museum covers an area of 37,000 square meters, of which the floor area is 18,000 square meters. The construction started in 1984 and the museum opened to public in November 1988. The exhibition building is a 4-storey castle-like construction. The exterior walls are covered with beige Taishan bricks. Two sides of the gate are decorated with a granite relief sculpture, *Map of Ancient Farming and Battling*. There is a pool in front of the gate. Lovely fish can swim in the clear water. 12 Ming Dynasty's sculptures of 12 symbolic animals of the years stand in the pool. Through transparent depolished glass, natural light enters the central hall with shale floor. Around the central hall, 32 galleries spiral upwards on 4 floors.

01

01. 馆舍前水池及明代十二生肖石雕

01.Pool in front of the exhibition building and the Ming Dynasty s sculptures of 12 symbolic animals of the years

02

02. 艺术自然馆与著名雕塑“闯”

03. 首层平面

04. 国内首次运用拉锁式玻璃幕墙技术的采光玻璃天棚

02.Shenzhen Museum of Art and Natural History and the famous sculpture “Opening up”

03. Plan of the ground floor

04. The glass ceiling adopts the technology of zipper-type glass curtain wall which is firstly created in China

03

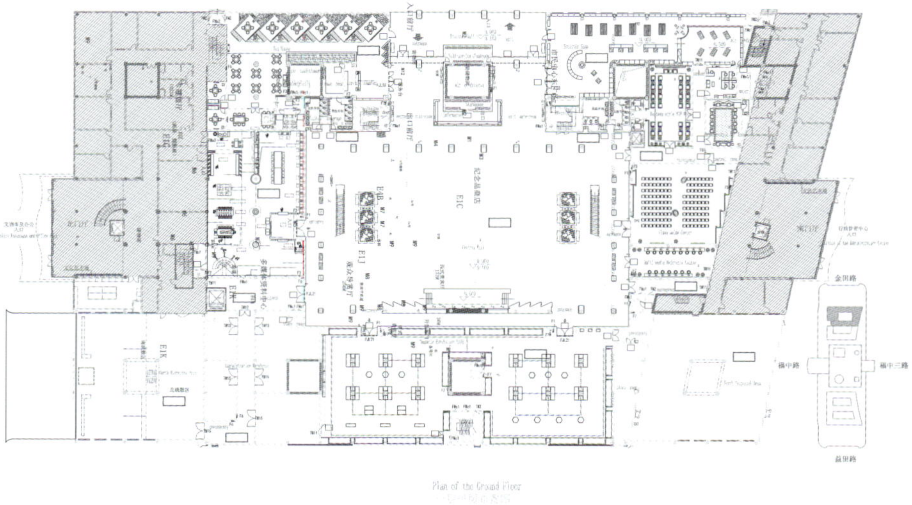

04

05

历史民俗馆位于深圳标志性大型建筑——“市民中心”东翼，整体造型气势磅礴。历史民俗馆建筑面积36 500 m²，2008年建成开馆。中央大堂面积2 160 m²，高23 m，庄重典雅，颇具气势。整个大厅和回廊采用色彩柔和、协调的大理石铺装地面与墙面，上方的采光玻璃天棚采用国内建筑中首创的拉锁式玻璃幕墙技术，通过智能遮阳系统，合理利用自然光，进一步烘托出建筑装饰色彩，让人感觉温馨舒适。正面装饰墙提取了极具岭南特色的窗格作为装饰元素，进一步突出了博物馆的文化氛围和典雅气质。中央大堂还独具匠心地设置了6个大型花坛，与整个中央大厅相得益彰。展厅共分为3层，为“回”字形全封闭结构，面积约9 000 m²。临时展厅面积1 100 m²，配备了先进的专业展柜和安防设备，可满足举办大型、高档次文物展览的需求。

Shenzhen Museum of History and Folk Culture is located in the east wing of the Citizens' Center, which is a symbolic landmark in Shenzhen. The museum covers a building area of 36,500 square meters, opened in 2008. The 2,160-square-meter central hall is 23 meters high, which creates a dignified and elegant atmosphere. The floor and walls of the central hall and corridors are covered with marble. The glass ceiling adopts the technology of zipper-type glass curtain wall which is firstly created in Chinese building industry. The intelligent sunshade system makes reasonable use of sunlight. The front wall of the central hall is decorated with window screens of Lingnan style, which makes the cultural atmosphere and elegant temperament of the museum more outstanding. There are 6 large flower-beds, which perfectly match the central hall. The galleries are distributed on three storeys, covering an area of about 9,000 square meters. The temporary gallery is about 1,100 square meters. Equipped with advanced showcases and security facilities, the gallery can meet the need of large and top grade exhibitions of cultural relics.

05. 历史民俗馆正门及东广场

对页

06. 历史民俗馆大堂

07. 多媒体资料中心

08. 儿童馆

05.Gate of Shenzhen Museum of History and Folk Culture and the east square

Opposite page

06.Central hall of Shenzhen museum of history and folk culture

07.Multi-media reference center

08.Children's hall

06

07

08

Guangxi Museum of Nationalities

广西民族博物馆

广西民族博物馆是一座以征集、收藏、保护、研究、展示广西12个民族繁衍生存、融合发展的文化与艺术遗存、物证、典籍等物质与非物质遗产为主的专题民族博物馆。博物馆于2006年3月开工建设，作为广西壮族自治区成立50周年之际的重大献礼工程于2008年12月竣工，2009年5月1日正式向公众免费开放。馆址位于南宁市青秀区南麓的邕江河畔，占地85 671 m²，总建筑面积约33 882 m²，总投资约2.3亿元。

广西民族博物馆主楼外形取材于富有广西地域和民族特色的铜鼓，以主楼为建筑中心，两条装饰着铜鼓鼓面的走廊连接大厅两翼，建筑后部为弧形的三层保护展示大楼。整个建筑犹如一只展翅飞翔的大鹏，遨游于青山绿水之间。整个馆区设有公共服务区、露天展示区、文物保护研究中心、业务与行政管理区、后勤服务区等5个功能区。内部设有多功能会议厅、专题图书馆、网上博物馆和青少年活动中心等。

Guangxi Museum of Nationalities (GMN) is mainly for investigating, collecting, preserving, studying and showing the materials, the cultural remains, the ancient codes and records, the intangible cultural heritages and their study findings concerning the reproduction, mixture and development of the 12 native ethnic groups in Guangxi Zhuang Autonomous Region. Its construction begun in March, 2006, and was completed in December, 2008 as a major project to celebrate the 50th anniversary of the establishment of Guangxi Zhuang Autonomous Region. It was opened to public in May, 2009. With Qingxiu mountain Forest Park to its north and Yongjiang River winding eastward to its south, GMN covers an area of 85,671 square meters, and a total floor area of 33,882 square meters. Its total investment is 23 million yuan.

The main building lies in the center, and it takes the shape of a bronze drum which represents ethnic culture in Gunagxi. The arc-shaped office building stands behind the main building, connected by two decorated corridors extended from both sides of the main building. Seem from distance, this architecture looks like a flying roc. The museum contains five functional area, including public service area, outdoor display area, conservation research center, business and administrative area, logistics service area. The main building is well equipped with a multifunctional conference room, a cultural relic sample observing room, a library, and a teenager club.

01

01. 露天展示园夜景

对页
02. 航拍
03. 露天展示园
04. 后勤管理楼

01. Nightscape of outdoor display area

Opposite page
02. Aerial view
03. Out door display area
04. Logistics Management Building

02

03

04

05

这是一座将铜鼓文化与现代建筑艺术高度结合的建筑。建筑外装饰采用现代建筑新型材料：金属铝板幕墙，中空 Low-e 玻璃幕墙及石材幕墙，铝锰镁金属屋面板。银白色的铝板拼接成铜鼓纹饰镶在蓝色玻璃幕墙表面，淡黄色麻面对缝石材外墙。颜色搭配协调合理、造型新颖轻盈。主入口的新型轻钢结构玻璃圆弧大雨棚，为游客提供了宽敞的集散地点，同时也为整个建筑主体增添雄大气势。

This architecture is a perfect combination of bronze drum culture and modern architecture technology. New modern architecture materials were used on the surface of the building, such as aluminium sheet panel wall, hollow Low-e glass panel wall, stone panel wall, and mental roof slab of aluminium, manganese, and magnesium. Wandering On the surface of blue glass curtain wall, are the bronze drum ornament made of silvery white aluminium plate. They match well in colour and shape. A wide arc canopy of new type light-weight steel structure was propped up at the entrance, which is broad enough for a large amount of visitors to gather.

05. 外景 05. Exterior

对页 **Opposite page**
06. 大堂 06. The lobby

Liuzhou Museum

柳州博物馆

柳州博物馆是一座拥有先进设施与现代化功能的综合性国家级重点博物馆，始建于1959年，现已逐步发展成为集文物保护、展示、研究、教育为一体的多功能现代化大型博物馆。

柳州博物馆建筑面积16 600 m^2，展示区面积6 800 m^2，建筑外观为古城阙样式，采用花岗岩风暴石砌筑，典雅厚重，外廊有21面浮雕，寓意21世纪的柳州和2 100年的厚重历史，反映柳州史前文明、历史文化、民族文化、现代文化、民间传说以及精品文物展示等丰富多彩的内容。

柳州博物馆主楼建筑共4层，1~3层为五大基本文物陈列馆。此外，还在一楼设多功能展示厅，在三楼设学术报告厅。整栋建筑具有完善的设施和管理服务功能。为照顾特殊参观人群，博物馆在出入口均设有轮椅坡道，内部垂直交通设置升降电梯，各层公共厕所内布置了无障碍厕位，博物馆整个公共场所均可无障碍到达。

Liuzhou Museum is a national key comprehensive museum with advanced facilities and modern functions. It was first built in 1959, has been developed into the center of preservation, display, research and education of cultural relics.

The floor area of Liuzhou Museum reaches to 16,600 square meters, exhibition area is 6,800 square meters. The building displays ancient city watchtowers and was built with granite, it's elegant and dignified. Its outer verandas are decorated with 21 relief sculptures which imply Liuzhou in the 21st century and 2,100 years of long-lasting history. They also reflect the prehistoric civilization, historical culture, nationality culture and modern culture of Liuzhou, folklore, cultural relics and so on.

Liuzhou Museum has four floors, the first, second and third floor are the exhibition areas of five basic galleries. Besides, the first floor has multi-functional hall and the third floor has the academic report hall, the whole building has perfect facilities and management services function. For service the disabled person, the building has wheelchair accesses, elevators and obstacle-free toilets, making sure that disabled person could go to everywhere in this building. As stated previously, the whole museum area has the leading concept design, functional facilities and perfect service.

01

01. 博物馆旧影

对页
02. 柳州博物馆
03. 总平面

01. Old photograph of the museum

Opposite page
02. Liuzhou museum
03. Master plan

02

柳州博物馆总体建筑设计布局科学合理、参观路线顺畅、重点突出，充分注重传统与现代的结合，体现人性化设计，达到人文与自然和谐统一，突出了文物性、文化性、民族性、功能性、特色性、公共空间性。此外，对于公共服务区等公共活动区域，充分利用自然光线照明和生态材料进行建筑设计，以“人文、低碳、自然、原生态”为理念，设计创造一个环境幽雅、品质高雅的高质量文化博览区域。

Liuzhou museum building has scientific and rational logistic topological design, the visitors' itinerary is unhindered. The design fully emphasizes the combine of tradition and modern, incarnates the humanization design and achieves the union of humanities and nature, highlights the characters of cultural relics, civilization, nationality, function, feature and public space. Besides, for the public activity space as public service area, the design makes the most of natural lights and ecological materials, follows the idea of “Humanities, Low Carbon, Nature and Original Ecology”, creates a public cultural exhibition area with graceful environment and High quality.

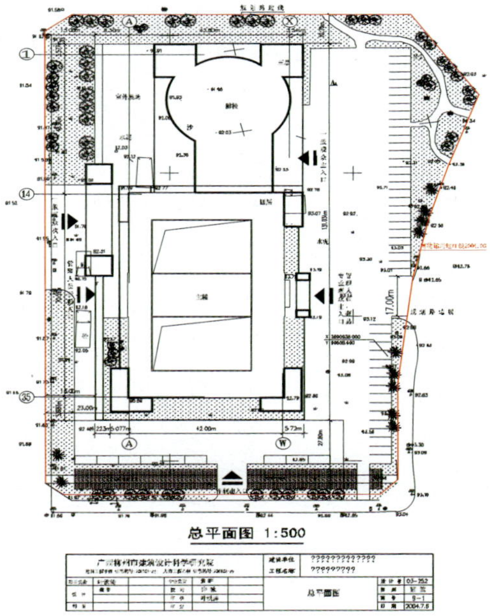

03

華天世紀酒店

04. 博物馆鸟瞰　04. Aerial view of the museum

05. “柳州历史文化陈列”——生动再现柳州 2100 多年悠久的历史和文化发展

06. “柳州民族风情陈列”——全方位了解柳州壮侗苗瑶等少数民族地区风情和民俗文化的窗口

05. Liuzhou History and Culture Exhibition——Reappear the 2,100-year history and culture of Liuzhou

06. Exhibition of Nationality Culture in Liuzhou——A window to fully understand the customs and culture of Zhung,Dong,Miao,Yao ?? nationalities areasin Liuzhou

The Museum of Hainan Province

海南省博物馆

海南省博物馆位于海口市国兴大道南侧，东临南渡河，西接城市公园，是海口文化公园的重要组成部分，建成后与海南省图书馆和海南省文化艺术中心共同组成海口文化公园的主体建筑，是海口市标志性建筑之一。

该馆的设计理念取自于“浮游之岛”。三条简洁的矩形形体平行展开，漂浮在水池之中，就像海南省地处南海之中一般，体现出海南省基本的地域特征。三个平展的形体表达了一种无限的延展性，象征着海南发展的无限广阔性。三个形体所产生的两个矩形虚空间又恰恰象征着无限的包容性，表达了海纳百川，吸引天下来客的意义。

水池中的水顺着虚空间一直延伸至博物馆内部，使博物馆形成了宁静、清新、高雅的气氛，同时又与文化公园整体规划相协调。木材、石材、钢材的运用，光的反射、折射与池中的倒影一起形成了历史与未来的对话。

设计将景观空间引入博物馆，阳光大厅、水景庭院、竹园与展厅和谐共生。视线可

Hainan Museum is situated on the south of Guoxing Road in Haikou, west to the Nandu River and east to the City Park. The museum is an essential part of the Haikou Culture Park, which also includes the Hainan Library and the Culture and Art Centre of Hainan. The museum will become one of the landmarks in Haikou when it is established.

The design of the museum gets its idea from the “floating island”. Three simple rectangular shapes floating in the pool are parallel to each other, which is just like the location of Hainan Province floating in the South Sea. This shows the basic features of Hainan's geography. While, the bodies of the three flat boxes with infinite ductility give us the imagination of the infinite development possibilities of Hainan in the future. At the mean time, two void spaces which are created by the three boxes represent the inclusive of the city and the sincere welcome to all the guests through out the world.

The water in the pool goes straightly into the interior of the museum along the void space, which not only attributes to the quiet and elegant atmosphere in the museum, but also gets coordinated to the general plan of the Culture Park. The use of materials such as wood, stone and steel, together with the reflection of light and shadows in the pool make the encounter of history and future coming true.

The design brings landscape into the museum where the exhibition space coexists harmoniously with the sunshine hall, water courtyard and

01

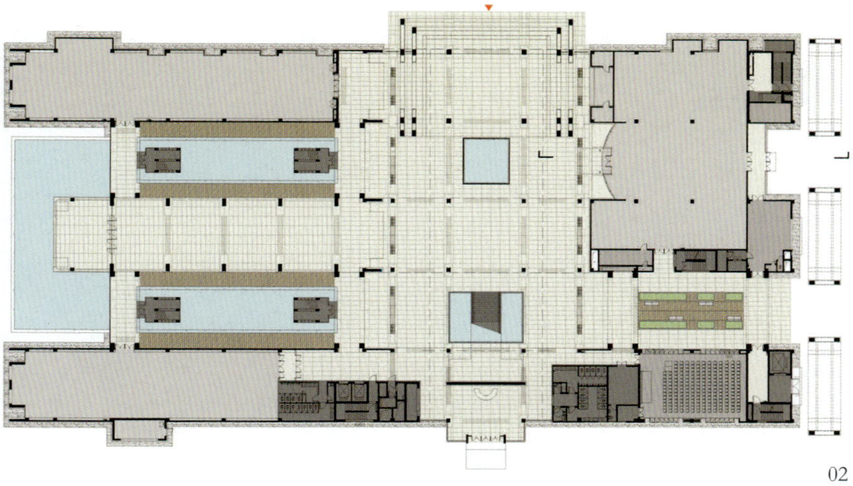

02

03

02. 一层平面
03. 立面

02. Plan of ground floor
03. Elevation

对页
01. 主入口夜景

Opposite page
01. Main entrance

04

05

06

04. 主入口
05. 局部立面
06. 东侧外观

04. Main entrance
05. Part of elevation
06. Appearance of east side

对页
07. 东侧立面
08. 西侧展厅外观

Opposite page
07. East elevation
08. Appearance of west hall

07

08

09

以通过开放的空间形成多种视廊，穿透博物馆厚重的墙体与文化公园相联系。共享大厅是博物馆的主要礼仪性空间，通透的空间、伞状的膜结构柱头、大面积木隔栅屋架以及透过屋架产生的丰富光影，都体现出浓郁的地域性。20 m高的大型浮雕墙，采用独具海南特色的纹样进行装饰，突出博物馆的历史感。

本项目充分考虑了海南热带地区的气候特征及文化背景，既体现出了博物馆建筑厚重的历史感，又在此基础上产生大量的灰空间与休闲场所，可以充分满足海口市民高层次的文化休闲需要。

the bamboo courtyard. Through the open space, sight line can penetrate the thick wall of the museum from variety of angles to have a connection with the Culture Park. The sharing hall is the main ritual space in the museum, which conveys the geographical traits sufficiently by the permeable space, umbrella-shaped membrane structure-chapiter, roof truss with large area of wooden grille and the shadow it casts. Large embossed wall of 20-meter-high which is decorated with the unique pattern of Hainan features highlights the history sense of the museum.

This project takes into consideration sufficiently of the climate features of Hainan in the tropics and the culture background of it, which shows the heavy sense of history of the building. On the basis of it, the design also creates lots of semi-public space and leisure space which will fulfill the need of citizens in Haikou to have higher level entertainment activities.

09. 城市门厅空间

对页
10. 展厅间室外水院

09. The hall

Opposite page
10.Water Court yard outside the exhibition hall

China Three Gorges Museum in Chongqing

重庆中国三峡博物馆

重庆中国三峡博物馆（重庆博物馆）前身为1951年成立的西南博物院。1955年6月因西南大区撤销，更名为重庆市博物馆，2000年9月国务院办公厅批准成立重庆中国三峡博物馆（重庆博物馆）。

重庆中国三峡博物馆于2005年6月18日正式对外开放。博物馆占地面积30 000 m^2，建筑面积45 098 m^2，展厅面积23 225 m^2。它位于最具重庆城市形象的心脏地块，与重庆人民大礼堂、重庆人民广场共同构成“三位一体”的城市标志性建筑群。博物馆主体建筑气势恢弘、内涵深邃，在外观设计上充分考虑与历史文化元素的完美融合：大面积的蓝色玻璃象征水，古朴的砂岩外墙代表山，充分体现出重庆山水之城的特色。进入博物馆内，观众便身处于三峡植物生态廊中，继续向前又豁然开朗。博物馆大厅以米黄、乳白色为基调，汉代石阙文物也陈列其中，自然光透过玻璃穹顶投射在大理石地面上，使室内空间更加宽敞明亮和简洁大气。观众怀着豁然舒畅的心情，与历史艺术直接“对话”。

重庆中国三峡博物馆现为首批国家一级博物馆、中央地方共建国家级博物馆培育对象、全国爱国主义教育示范基地、全国青少年教育基地和重庆市科普基地。

China Three Gorges Museum in Chongqing was founded in 1951 as Southwest Museum. It was renamed in June 1955 as Chongqing Museum since the withdrawal of the Great Southwest District, and again in September 2000 as China Three Gorges Museum in Chongqing (Chongqing Museum) with the permission of the General Office of the State Council of China.

China Three Gorges Museum in Chongqing was formally opened to the public on June 18th, 2005. It covers an ares of 30,000 square meters. Its floor area counts to 45,098 square meters, among which 23,225 square meters is exhibition halls. The museum lies in the heart of Chongqing city, forming the city's landmark together with the People's Great Hall of Chongqing and the People's Plaza of Chongqing. The main building of the museum which intends to embody local history and culture and present the landscape characteristics of the city is splendour. Entering the museum, visitors are welcomed immediately by a porch decorated with plants from the Three Gorges area. Different shades of cream set the tune for the atrium where the stone Que of the Han dynasty are displayed. The marbled floor is enlightened by the natural light coming through the glass dome, which accentuates the brightness and elegance of the space. From here, visitors set off their amazing journey around the museum, communicating with the past.

The museum has been one of the earliest national museums. It now receives policy and financial support from both central and local governments, and is selected as the national patriotism education demonstration aase, the national juvenile education base and the public scientific education base of Chongqing.

对页
01. 外景
02. 全景

Opposite page
01. Exterior
02. Panorama

03

04

03. 博物馆大厅
04. 重庆城市之路展厅

对页
05. 抗战岁月展厅
06. 大三峡数字环幕电影
07. 壮丽三峡展厅
08. 远古巴渝展厅

03. Hall of the museum
04. Chongqing History Hall

Opposite page
05. Anti-Japanese War exhibition hall
06. Big Three Gorges – digital circular – screen movie
07. The Magnificent Three Gorges exhibition hall
08. The Ancient Chongqing exhibition hall

05

06

08

07

Sanxingdui Museum

广汉三星堆博物馆

三星堆博物馆位于全国重点文物保护单位三星堆遗址东北角，地处历史文化名城广汉城西鸭子河畔，南距成都 40 km，北距德阳 20 km，是我国一座现代化的专题性遗址博物馆。馆区占地面积约 400 200 m^2，建筑总面积约 18 000 m^2，绿化面积达 80% 以上，环境优美，布局动静得宜，现代园林景观与周围自然环境融为一体，充分体现了“文、秀、清、幽”的川西园林之美。

园区主要建筑包括三星堆博物馆青铜馆、综合馆、游客接待中心、购物中心、文保中心及其他文化配套设施。主要建筑从外观造型到内部陈列都突破了传统历史博物馆的格局，追求与地貌、史迹及文物造型艺术相结合的神韵，融原始意味和现代气息为一体，力图表现三星堆文化的苍古雄浑及三星堆文明的博大精深。

Sanxingdui Museum, a modern special topic site museum, in the northeast of Sanxingdui site, an important heritage site under state protection, is located on the bank of Yazi River in the west of Guanghan City which is a historical cultural city. It is 40 kilometers to the north of Chengdu City and 20 kilometers to the south of Deyang City. The museum covers an area of about 400,200 square meters with a floor area of about 18,000 square meters and a green area of over 80% in excellent surroundings. With an appropriate combination of dynamics and statics and a harmonious entity of modern garden landscape and the surrounding natural environment, the museum fully embodies the beauty of gardens in Western Sichuan in terms of “Culture, Elegance, Tranquility and Peace”.

The site mainly comprises of Bronze Hall, Comprehensive Hall, Visitor Reception Center, Shopping Center, Cultural Relic Protection Center as well as other cultural supporting facilities. The main buildings make a breakthrough in patterns of the traditional historical museum in terms of the exterior styling and interior display; seek the romantic charm of integration with the land feature, historical sites and the formative arts

01

01. 博物馆晚霞

对页
02-03. 外景

01. The Museum in Sunset

Opposite page
02-03. Exterior

02

03

04

三星堆博物馆青铜馆建筑面积 7000 m^2，馆体外形为变形螺旋式建筑，建筑造型奇特，风格古朴凝重，强调“堆”的形象、突出自由攀升的螺旋曲线，具有“堆列三星”和“人类历史演进历程”的双重寓意，印证了“建筑本身应像土里面生出来的”的格言。综合馆建筑面积约 5 000 m^2，外形为半弧形斜坡生态掩土型建筑。其主体覆以草坪，外部近侧辅以山石飞瀑与多种常绿植物，建筑外观与周边园林设计两相浑融，意在表达人与自然和谐共生的人文精神。两大展馆陈列规模宏大，布局考究，匠心巧运，融知识性、故事性、观赏性、趣味性于一炉，有力地揭示出三星堆文物的深刻内涵，集中反映了三星堆文明的辉煌灿烂，清晰地勾勒出古蜀历史文化的发展脉络，给人以身临其境、故国神游之感。

of cultural relics; blend the primitive tincture with modern touch to display the primitive and vigorous Sanxingdui Culture and the significant and profound Sanxingdui Civilization to the utmost extent.

With an floor area of about 7,000 square meters, Bronze Hall in Sanxingdui Museum is characterized by a unique spiral outline with a simple and dignified style, emphasizing the image of "Pile-up" and highlighting a spiral curve rising freely with a double implied meaning of "Piled-up Sanxingdui Site" and "Evolutionary Process of Human History", which just echoes the old saying "The building should be just seemed to be grown from the soil". The Comprehensive Hall with an ecological-earth-sheltering-half-arch-slope outline enjoys an area of about 5,000 square meters. Its main part is covered with grass lawn and the surroundings are provided with mountains, stones, waterfalls and various evergreen plants. The exterior of the building and the surrounding garden design are blending harmoniously with the purpose of displaying the humanistic spirit of the harmonious coexistence of human and nature. The two large-scale and exquisite exhibition halls, with an integration of knowledge, tales, appreciation and interest, show the profound meaning of Sanxingdui artifacts in a convincing way; reflect the splendidness of Sanxingdui Culture concentratedly;

04. 综合馆

对页
05. 博物馆大门

04. Complex exhibition hall

Opposite page
05. Gate of the museum

06

2000 年 2 月 24 日，胡锦涛同志参观三星堆博物馆后，对博物馆建筑给予高度评价："这个博物馆设计得不错，有新意、有特色，螺旋式上升是人类文明发展的轨迹。

draw the development process of the ancient Shu in terms of history and culture, all of which give people a feeling of personally visiting the ancient Kingdom.

"The design of this museum is quite pretty with innovations and distinguishing features; the spiral ascending is just the development track of human civilization." Acclaimed by Hu Jintao, the President in China, after his visit of Sanxingdui Museum on Feb. 24, 2000.

06. 馆园
07-08. 展陈大厅

06. Garden
07-08. Exhibition hall

07

08

Dengxiaoping's Museum

邓小平故居陈列馆

邓小平故居陈列馆位于四川省广安市广安区协兴镇牌坊村，距邓小平故居约 500 m，在邓小平诞辰 100 周年之际落成开馆，江泽民题写馆名。

陈列馆运用国际博物馆展示的成功理念，结合现代科学技术的先进展示手段，主题鲜明，特点突出，内容丰富，馆藏珍贵，设计新颖，庄重大方，是目前国内第一家以纪念邓小平为专题的博物馆。

陈列馆建筑采用钢筋混凝土框架结构，建筑面积 3 800 m^2。它以精炼简朴的建筑语言，在现代建筑设计理念中融入川东民居的建筑风格，反映了中国改革开放历史性变化的深刻内涵。

陈列馆坐西向东，一字排开，三个青瓦坡形屋面，三叠三起，一起比一起高，最后耸立起一座丰碑，寓意邓小平“三落三起”的传奇人生。

陈列馆背靠绵绵小山，面向广袤大地，绿树鲜花环绕，碧水倒影婆娑，既有广安的地域特色，又反映了时代精神，与周围环境紧密结合，美轮美奂，风格十分独特。

陈列馆由序厅、三个陈列展厅、电影放映厅、珍藏陈列厅组成。陈列布展工作是根据中共中央关于邓小平同志诞辰 100 周年纪念活动的通知精神，在四川省委的直接领导下，在中共中央文献研究室的协助下，在邓小平同志亲属的大力支持下进行的。

The Museum of Dengxiaoping's former residence is located in Xiexing town of Guangan city ,Sichuan province. It is 500 meters away from Deng's old courtyard . The museum is open to the public at Deng' 100th birthday in August of 2004. Jiangzeming inscribes on the board with Dengxiaoping' s Museum.

Deng's Museum has a construction area of 3,800 square meters. The scene is very beautiful with verdant trees around. The combination of traditional and modern ideas is adopted to reflect Deng's extraordinary life. The museum looks eastern. In the right , the three sloping roofs represent Dengxiaoping's experience of three times up and down. In the middle , the high rising building is called Monument to represent his great achievement and summon us to lift great banner of Dengxiaoping's theory.

Deng's Museum consists of a Preface hall , three exhibition hall , a cinema and a treasure hall. It is the only one museum to memorize Dengxiaoping in China at present. Under the leadership of Sichuan province of CPC , professors of Document Research Centre of Central Committee and Dengxiaoping's relatives have supported greatly this exhibition. Hi-tech sound, lighting and electronic devices are used in the exhibition hall to achieve a better presentation of Deng's great life.

对页
01. 陈列馆全景
02. 陈列馆序厅
03. 总平面

Opposite page
01. Panorama
02. Prelude Hall
03. Master Plan

01

03

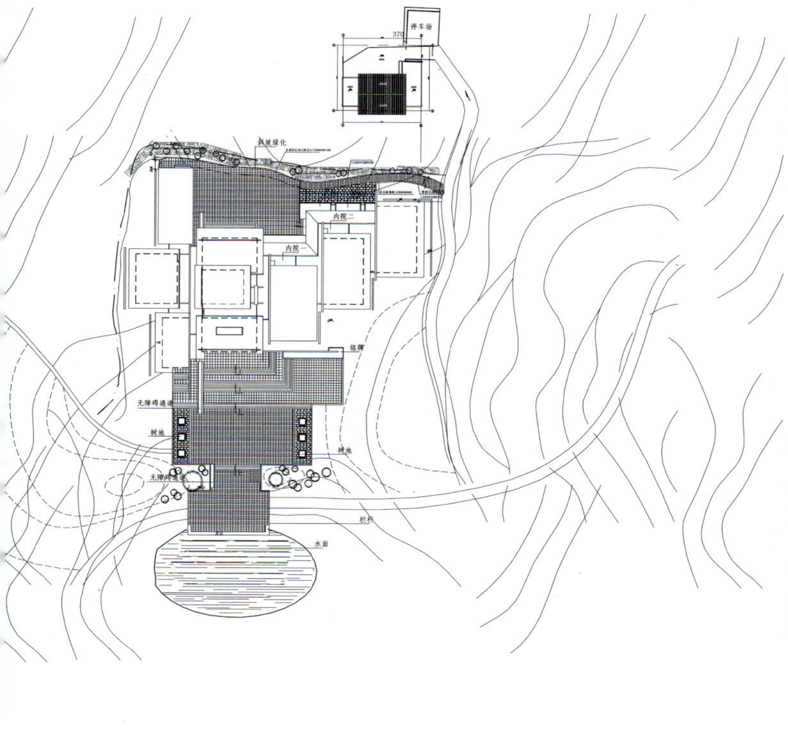

02

戎马生涯
MILITARY LIFE
八路军一二九师师部
邓小平住所旧址

04

"邓小平小道"
DENG XIAOPING PATH
开创伟业
FOUNDING GREAT CAUSE

05

06

07

06. 陈列馆三展厅
07. 陈列馆珍藏厅

对页
04. 陈列馆一展厅
05. 陈列馆二展厅

06. 3rd Exhibition Hall
07. Hall of Valuable Exhibits

Opposite page
04. 1st Exhibition Hall
05. 2nd Exhibition Hall

Jinsha Site Museum
金沙遗址博物馆

金沙遗址博物馆位于成都市市区西部，在2001年2月发现的金沙遗址原址上建立，是一座为保护、研究、展示金沙遗址及古蜀文明而设立的专题博物馆。占地面积30万 m²，由遗迹馆、陈列馆、文物保护与修复中心、文化交流中心、游客接待中心等部分组成。总建筑面积38 000m²，绿化面积138 143m²，建筑容积率0.15%，绿地率70.7%。

馆区内以横贯东西的摸底河为横向景观轴，以南北轴线的开放空间形成纵向文化轴，合理地功能布局，充分实现了现代博物馆教育、休憩、观光、游览的各种功能。

博物馆内两个主体建筑遗迹馆和陈列馆，一方一圆，均采用斜坡状的现代化全钢架结构建筑。建筑与地表斜坡连接的方式，减弱了建筑体量，强调了博物馆建筑与遗址的紧密关系。

In the northwest of Chengdu, there stands a thematic museum--Jinsha Site Museum, which is established in situ, discovered in February 2001, with the purpose of preservation, research, exhibition of Jinsha site and ancient Shu culture. This museum covers an area of 300,000 sq.m., it consists of the excavation hall, the exhibition hall, preservation and restoration center of cultural relics, culture exchange center, and tourists center etc., with total construction area 38,000 sq.m. and green coverage more than 138,143 sq.m. (floor area ratio of 0.15%; greening rate of 70.7%).

The museum horizontal axis is landscaped along the Modi River from east to west, while its vertical axis is cultural exhibition on an open space from north to south. The proper functional layout fully achieves various functions of a modern museum, which are education, resting and sightseeing.

The main constructions of the museum are the excavation hall with round style and the exhibition hall with square style both adopting modern sloping steel-framed structure. The design of sloping-roof construction extending to the ground is to weaken construction volume, emphasizing the close relationship between the museum and the site.

01

01. 中国文化遗产标志——太阳神鸟永久纪念雕塑

对页
02. 博物馆鸟瞰
03. 总平面
04. 夕阳映照下的博物馆南大门

01. Sculpture of Sun-Bird, Symbol of Chinese Cultural Heritage

Opposite page
02. Aerial view of the museum
03. Master plan
04. South gate of the museum in sunset

02

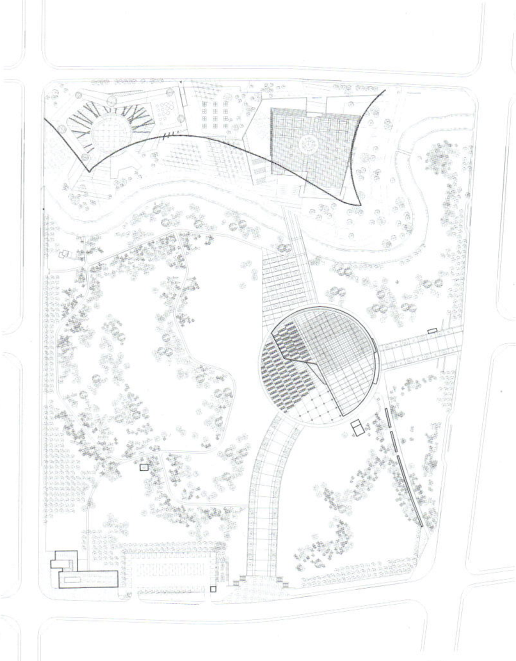

04

总平面 Master plan

03

05

06

金沙遗址博物馆建筑与遗址环境的高度协调，确保了博物馆建设的整体效果，体现了遗存本体及其历史环境的完整性、持续性，可持续发展的规划理念，也为日后的科研、展示、教育等功能发挥预留下了足够的发展空间。大气恢弘的现代化建筑设计更受到各级领导、国内外专家及观众的一致好评，荣获了 2008 中国建筑师学会建筑创作金奖。

The perfect harmony of museum constructions and the site environment guarantees the overall effect of the museum infrastructure, reflecting integrity and sustainability of original site and historical situation. Sustainable development planning principle offers great potential development for future scientific researches, exhibition and education, etc. This magnificent modern architecture was unanimously praised by museum directors, experts and visitors has winned Gold Medal on Architecture Creation Awarded by Chinese Institute of Architects in 2008.

07. 遗迹馆外墙夜景

对页
05. 遗迹馆外景
06. 原状展示的考古发掘现场

07. Nightscape of Relic Hall's facade

Opposite page
05. Exterior of Relic hall
06. Site of archaeological excavation in original state

08

08. 陈列馆外景
09. 陈列馆宽敞适宜的公共空间

08. Exterior of Exhibition Hall
09. Broad and Comfortable Public Space

对页
10. 陈列馆内随光影变化的的太阳神鸟穹顶

Opposite page
10. Sun-Bird Ceiling in Exhibition Hall Varied with Sunlight

09

Jianchuan Museum, China

建川博物馆

建川博物馆，全称建川博物馆聚落(Jianchuan Museum Cluster)，坐落于国家级历史文化名镇四川成都安仁古镇，占地 500 亩 (333 500 m²)，现已建成开放 15 座博物馆，共收藏文物 800 余万件，其中国家一级文物 121 件。为中国最大的博物馆聚落。建川博物馆由抗战、"红色年代"、地震、民俗四大主题博物馆组成，其中抗战博物馆系列包括正面战场馆、不屈战俘馆、川军抗战馆等分馆及抗战老兵手印广场和中国壮士群塑广场。旨在为了和平，收藏战争；为了未来，收藏教训；为了安宁，收藏灾难；为了传承，收藏民俗。该馆是目前国内民间资金投入最多、建设规模和展览面积最大、收藏内容最丰的民间博物馆。

建川博物馆聚落现馆藏珍贵文物超过 1 000 件，其中国家一级文物达 121 件，这在国内民营博物馆中首屈一指。建川博物馆聚落匠心独具地突破了传统意义上的单纯的"博物馆"的概念，不仅超乎想象地在国内第一次将多达 20 余个博物馆汇集在一起，而且还进一步将各种业态的配套如酒店、客栈、茶馆、文物商店等各种商业等汇集在一起，让这些配套设施呈现亚博物馆状态，形成一个集藏品展示、教育研究、旅游休闲、收藏交流、艺术博览、影视拍摄等多项功能为一体的新概念博物馆和中国百年文博旅游及乡村休闲度假旅游目的地。

Jianchuan Museum, also named Jianchuan Museum Cluster, is located in Anren, a national-class historical and cultural town in Chengdu, Sichuan. Covering an area of 500 mu (333,500 square meters), the museum cluster is made up of 15 museums. It has a rich repository of over 8 million pieces of artifacts, with 121 of them classified as Class-A National Treasures. This is the largest museum cluster in China. The museums are organized by four major themes: Second Sino-Japanese War (1937–1945), the 'Red era', the Wenchuan Earthquake, and Folklore & Culture. We exhibit items from the war to promote peace; present teachings from the past to develop a better future; recreate scenes of natural disasters to foster a sense of tranquility; collect items of folklore and traditions to reflect the soul of the people. Jianchuan Museum Cluster is the folk museum which is at present the domestic private investment, construction of the scale and exhibition area of the largest and contents of the most abundant.

The Jianchuan Museum Cluster has a rich repository of over 10 million pieces of artifacts, with 121 of them classified as Class-One National Treasures, which is number one in domestic private museums. The cluster espouses the philosophy of "viewing artifacts at leisure, thus appeasing both mind and body". It gets down to the broken ends of the traditional sense of the concept of "museum". It is not only beyond imagination at the first time gathering as many as 20 museums, but also supporting hotels, teahouse, antique shops and various commercial facilities. The necessary facilities match up with the cluster. It shows the new concept for museum, with the function of a set of display, education research, travel and leisure, collection and communication, art visit, cinema arts and so on. It will be the cultural tourism for centuries and the leisure vacation travel destination of China.

01

01. 雕塑

对页
02. 正面战场馆
03. 川军抗战馆

01. Sculpture

Opposite page
02. Frontal battlefield hall
03. Sichuan resistance army hall

抗战文物陈列·正面战场
Exhibition of Cultural Artifacts from China's Resistance War
against Japanese Aggression·The Hall of the Frontal Battlefield

02

抗战文物陈列·川军抗战
Exhibition of Cultural Artifacts from China's Resistance War against Japanese
Aggression·The Hall of the Sichuan Army in the War of Resistance

03

04

05

06

07. 不屈战俘馆出口

对页
04. 不屈战俘馆外景
05. 不屈战俘馆主入口
06. 不屈战俘馆内景

07. Exit of Unyielding War Prisoner Hall

Opposite page
04. Exterior of Unyielding War Prisoner Hall
05. Main entrance of Unyielding War Prisoner Hall
06. Interior of Unyielding War Prisoner Hall

Guizhou Museum

贵州省博物馆新馆

贵州省博物馆新馆位于贵阳市金阳新区迎宾路南侧，市行政中心东南侧观山公园东北角地块，西南临观山湖。新馆占地 47 400 m^2，总建筑面积 42 996 m^2，其中陈列展览用房 16 610 m^2，文物库房及文物库房管理用房 9 500 m^2，教育及综合服务用房 3 280 m^2。

贵州省博物馆新馆的设计构思理念为："菱石虽小，内蕴贵州文化千山千岛"，"织锦龙纹，解码夜郎人群遗传基因"。其外观来自多元一体的菱石理念，似菱形巨石被切削，是坚固永恒的精神象征。

博物馆功能的革新使贵州省博物馆新馆从单一的收藏展览功能转变为社会活动中心。为此将其构架和形态相应转变，新馆复杂的活动功能被压缩、重组为七大功能板块，其中收藏、展示、停车和管理四大组团为固定板块，而城市客厅及礼仪大厅、教育与培训、消费与公众服务三大组团为活动板块。这些组团通过室内的街道和室外的"非遗栈道"相连，形成一个吸纳环境能量的有机共生的未来博物馆。

The New Department of the Guizhou Museum situated on the south of Yingbin Road, Jinyang New District in Guiyang. It locates to the southeast of the administration center of the city, northeast of the Guanshan Park and southwest of the Guanshan Lake. The site covers an area of 47,400 square meters and the overall area of the new building is 42,996 square meters,which includes 16,610 square meters for display and exhibition, 9,500 square meters for conservatory and management, 3,280 square meters for education and service.

The aspiration of the project can be explained as "little stone contains mountains of profound culture of Guizhou", or "brocade with dragon pattern decodes the genes of the Yelang People "and so on. The outlook of the building is just like a diamond huge stone which is chopped and represents the eternal and unshakable spirit.

The innovation of the function makes the museum not only collecting and exhibiting objects, but also becoming the centre of society activities. In order to adapt to the new change, the complicated function of it was compressed and reorganized as seven parts, of which collecting, exhibiting, parking and management are the regular parts, while the urban and ritual hall, education and training, and public service are the temporary parts. All of these parts are connected by streets exterior and interior, which constitute an organic future museum absorbing energy from the environment.

The museum uses the technology of the operculum-like construction of the external wall, which makes the wall to have the ecology function of breath. Moreover, LED photodiode was buried in the wall to make the skin able to have different light effects and switch among them. The stone material used in the external wall was cut by

对页
01. 外景效果图
02. 总平面
03. 鸟瞰效果图

Opposite page
01. Exterior rendering
02. Master plan
03. Aerial view rendering

01

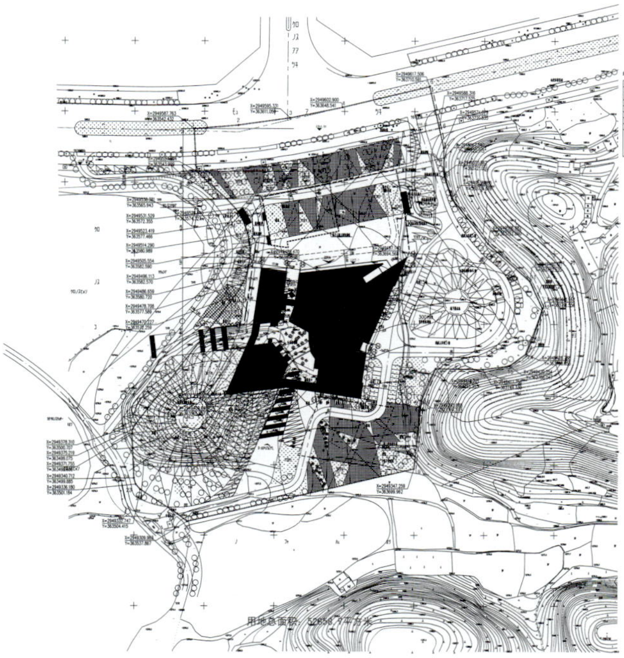

03

02

贵州省博物馆新馆采用了鱼鳃式外墙构造，新馆表皮可行使生态呼吸功能；在外墙中嵌入LED光电二极管灯，外皮可实现灯光效果的变幻和控制；外墙石材采用图案化电脑辅助切割，贵州自然历史文化的图像信息可以保存在外墙中供人发掘。

patterning computer's assistant, Thanks to that, the image information of natural and historian culture of Guizhou will be saved in the wall and is ready for the visitors to discover.

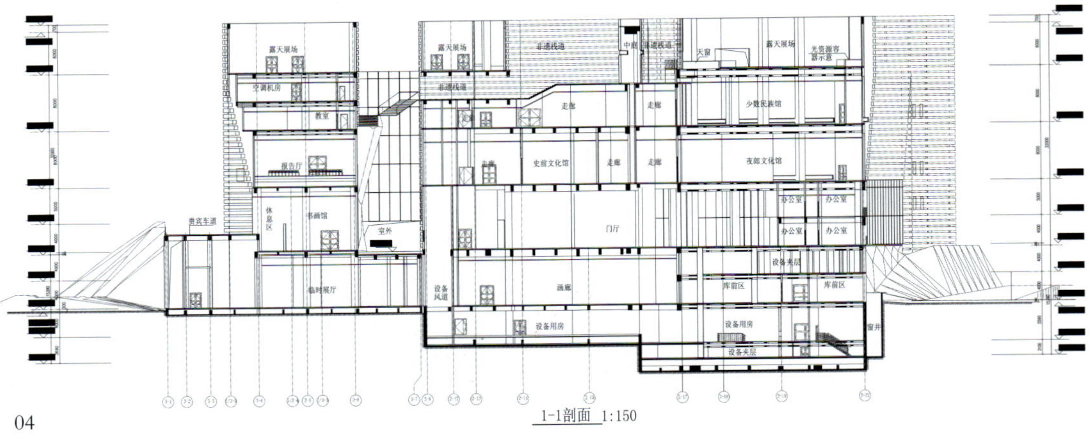

04

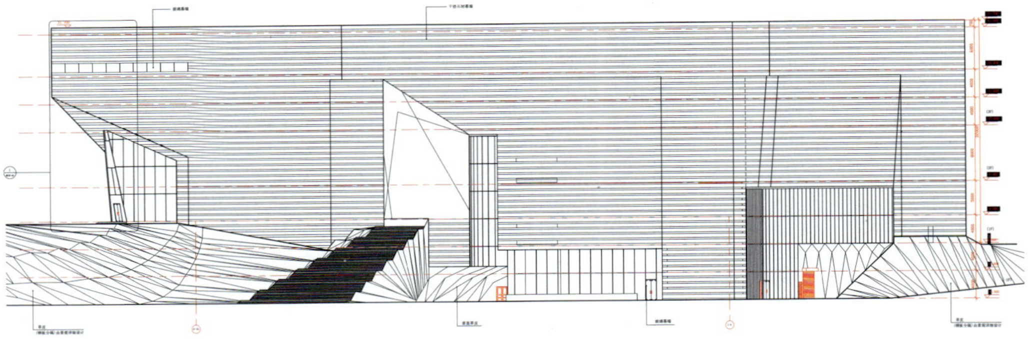

05

04. 剖面	04. Section
05. 立面	05. Elevation
06. 外景效果图	06. Exterior rendering
对页	**Opposite page**
07. 中庭	07. Atrium

06

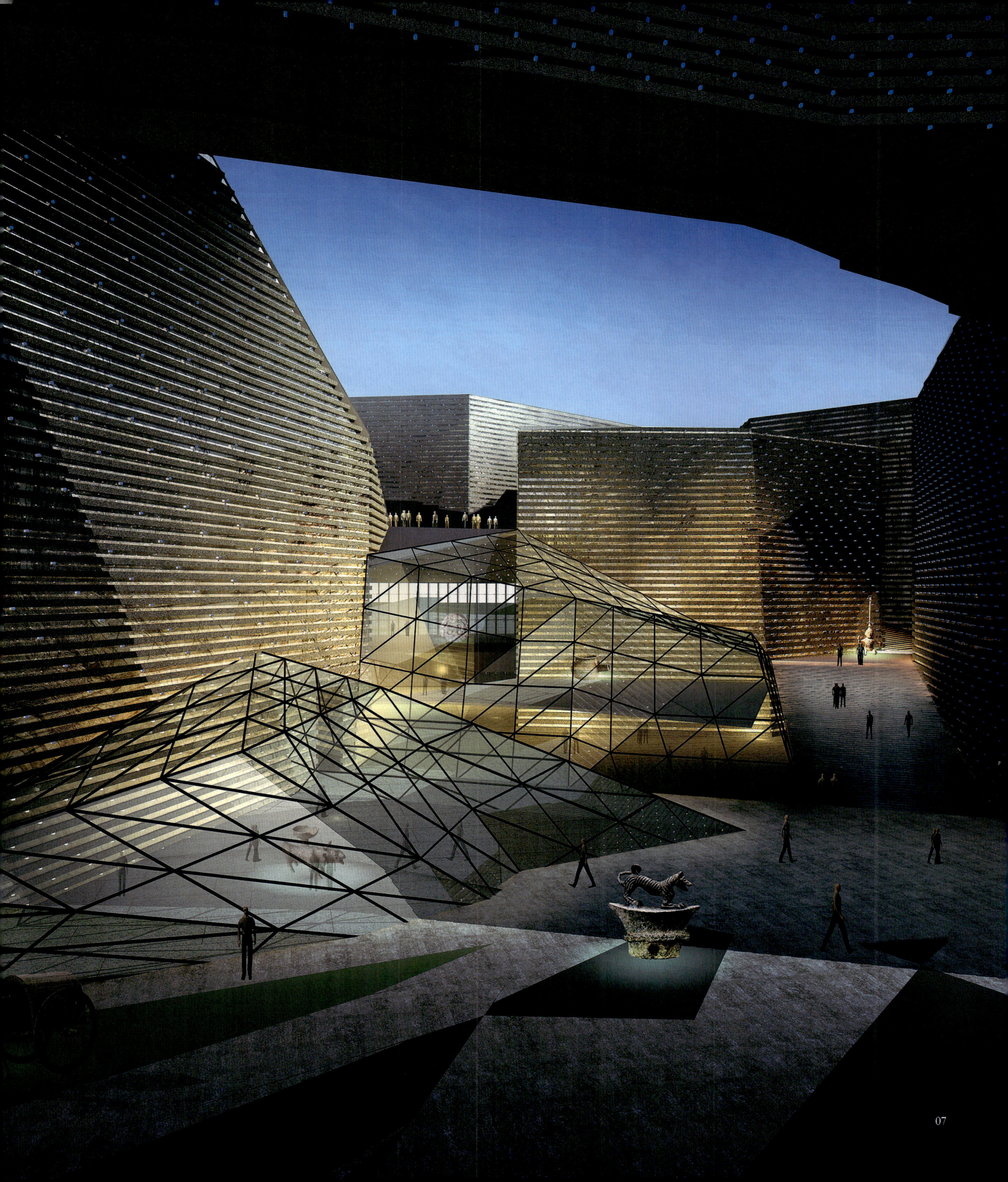

The Museum of Tibet

西藏博物馆

西藏博物馆位于拉萨市民族南路 2 号，是西藏有史以来的第一座现代化博物馆，由国家投资近 1 亿元兴建，1999 年 10 月 5 日正式开放。西藏博物馆占地面积 53959 m^2，建筑面积 23508 m^2，展厅面积 10451 m^2，由主馆区、办公生活区、民族文化苑三部分组成，其建筑规模宏大，民族特色浓郁，环境优美，是拉萨著名的标志性建筑和重要的旅游景点之一。

西藏博物馆馆藏文物丰富，特色浓郁，几年来始终坚持收藏保护、学术研究和展示宣传相结合的方式，发挥文物荟萃的优势，上穷远古先民的鬼斧神工，下含近现代名家的精品佳作，展现了不同时代文化艺术的精品，可使参观者领略到历史文化殿堂的博大与精深。西藏博物馆推出的基本陈列“西藏历史文化”以史前文化、不可分割的历史、文化与艺术、民俗文化四个部分为主要内容。此外，还设有“藏北自然资源专题展”、“明清瓷器精品馆”、“元、明、清玉器精品馆”、“叶星生民间文化捐赠馆”等各具特色的专题陈列展览。每个展馆都经过精心的设计和制作，风格各异，突出特色，集中展现了藏民族文化的魅力。

The museum of Tibet situated on the Minzunan Road, NO.2 in Lhasa, which is the first modern museum in the history of Tibet. The museum was sponsored by the government of 100,000 yuan and was officially opened on October 5th, 1999. The overall area of the site is 53,959 square meters, the floor area of 23,508 square meters and the exhibition area is 10,451 square meters. The museum separates itself into three parts as the main area, the working and living area and the national culture area. It is one of the remarkable buildings and important sights in Tibet, which is famous for its strong racial features and beautiful environment.

The museum of Tibet has acquired a fine collection of relics with remarkable characteristics. Over the years, the museum sticks onto the discipline of integrating collection, researching and display together, which make the visitors be fully aware of the vast and extensive culture in the history. The theme of the regular display is “History and Culture of Tibet”. The display contains four major parts that is prehistoric culture, inalienable history, culture and art, and folk culture. Besides, there is also some special topic exhibition such as“natural resources in the north of Tibet”,“elaborate works of porcelain of Ming and Qing Dynasty”,“fine works of jade of Yuan and Ming and Qing Dynasty”, “the donation of Ye Xingsheng of folk culture” and so on. Each of them was designed and arranged elaborately and shows the visitors various characteristics and charm of the Tibetan culture.

01

01. 历史照片

对页
02. 博物馆全景
03. 民俗苑文化长廊

01. Historical photo

Opposite page
02. Panorama of the museum
03. Gallery of folk’s culture

02

03

Shaanxi History Museum

陕西历史博物馆

陕西历史博物馆是按“轴线对称、主从有序、中央殿堂、四隅崇楼”的宫殿庭院式布局建造的高台基群体组合建筑。项目占地将近 70 000 m²，建筑面积 55 600 m²。中轴线上的主体建筑南北两面均为四阿顶重檐建筑，东西两翼为“吕”字形建筑，四隅各有崇楼一座。连廊把主体建筑、两翼建筑和四隅崇楼连接一起，使这座建筑主次分明、散中有聚、恢弘大气，恰到好处地表现了中国传统宫殿建筑“太极中央，四面八方”的空间构图特色以及千百年来早已潜入中国人空间意识中的“超以象外，得于寰中”的东方宇宙哲理。

National Museum of the Shaanxi History is consist of buildings which is in the style of Tang dynasty architecture with overall layout of the building looking like a palace with halls in the centre and elegant pavilions standing in the four corners. It covers an area of 70,000 square meters and building area of 55, 600 square meters. The main buildings on the central axis are layered eaves in the south and north, the east and west are designed with buildings in style of the Chinese character “吕” and the four corners with four tower buildings in the four corners. The corridor connects the main buildings, buildings on either side, and the four tower buildings, which make the buildings transparent and elegant. Thanks to the traditional design, the buildings are of the characteristics of “central taiji and all sides” and the universal philosophical truth of the oriental people.

01

02

03. 博物馆东南角全景鸟瞰
04. 序言大厅

对页
01. 原陕西省博物馆（摄于 1958 年）
02. 博物馆外景

03. Aerial view panorama of the muesum's southeast corner.
04. Foreword Hall

Opposite page
01. Historical photo of the museum(1958)
02. Exterior of the museum

Terra-cotta Warriors and Horses Museum

秦始皇兵马俑博物馆

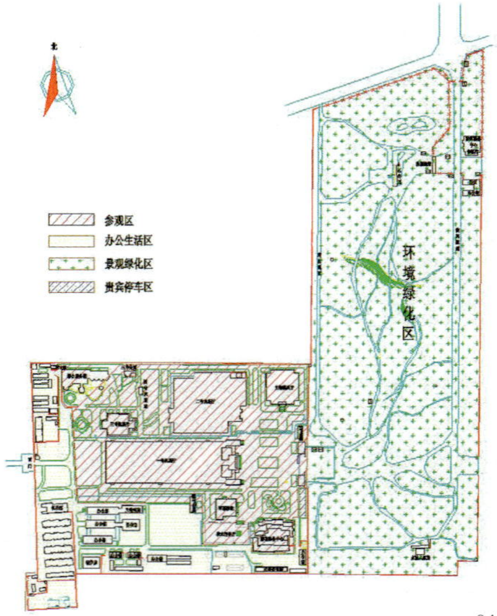

01

02

01. 博物馆总平面
02. 博物馆全景

对页
03. 秦俑一号坑内视全景

01. Master plan
02.Terra-Cotta warriors and horses museum mapping

Opposite page
03. Pit NO.1 inside view

14
一号坑的建筑
The Construction of Pit 1
秦俑的雕塑
The Sculpture of the Terra-cotta Warriors

03

秦始皇兵马俑博物馆是在秦始皇陵兵马俑陪葬坑遗址上建立的大型遗址性专题博物馆。馆区以一、二、三号坑保护展示厅、秦始皇陵出土文物陈列厅、秦文化学术报告厅、游客中心等建筑和游客集散广场及全方位绿化区域组成。根据文物遗址分布的特点，建筑总体采用了非对称布局，并在建筑体量、高度、色调、选材上适当减弱，作为对遗址从属关系的处理。建筑格调浑厚、古朴、粗犷、有力，与环境谐调统一。同时又考虑和秦陵展示的关系，三个主展厅（一、二、三号坑）东西向均作为观众的主入口。

秦兵马俑一号坑保护展示厅为落地式三铰拱架结构。展厅共设落地式三铰钢拱架18榀，东西长204.5 m，南北宽72.5 m，拱架弦长67 m，中矢高13.4 m，建筑面积15 458 m^2。展厅四周设高侧钢窗134樘，外墙面为普通水刷石墙面。

秦兵马俑二号坑遗址保护展示厅位于一号坑展示厅北侧东段20 m处，平面略呈曲尺形，东西全长134.20 m，南北全长106.25 m，建筑面积17 934 m^2。

The Terra-cotta Warriors and Horses Museum is a large special remain museum built on the Emperor Qin Shi Huang' s burial pits. The museum is made up of three pits, Qin Mausoleum exhibition hall, Qin culture report hall, tourist center and square, planting area. The museum take non-symmety arrangement according to the relics remain distribution character to decrease the volume, height, colour and material as the subordinate relation of the remain. The museum is of broad, classical, steady, harmony style. Three pits take the entrance of the east and west direction related to the Qin Mausoleum.

Pit NO.1 of Terra-Cotta Museum is a straight triangle steel structure of the total number 18 pieces, they are 204.5 meters long, 72.5 meters wide. 67 meters long of the arch, 13.4 meters high in the middle, covered the area of

04

05

秦兵马俑三号坑遗址保护展示厅位于一号坑展示厅北侧西段 12 m 处，平面呈方形，建筑面积 1 694 m^2，垂直高 6.2 m。主厅结构形式和建筑外观基本同二号坑遗址保护展示厅相似。

秦始皇陵出土文物陈列厅位于一号坑前广场东北侧，展厅平面略呈方形，建筑外形采用汉代坞堡形式，框架结构，建筑高 15.4 m，面积 7 231 m^2。

15 458 square meters. 134 pieces of high steel windows around, the out part is water brushed wall.

Pit NO.2 is located 20 meters north east of pit NO.1 of the square structure, 134.2 meters long east to west, 106.25 meters long north to south, totally 17 934 square meters of the whole area.

Pit NO.3 is located 12 meters west of NO.1 of square shape, covered the area of 1 694 square meters, 6.2 meters in height. The building structure and shape is similar to pit NO.2.

Qin Shi Huang' s Mausoleum relics exhibition hall is located north east of visiting area in front of pit NO.1 of the square shape just like the castle in Han dynast in frame work,15.4 meters in height, totally 7 231 square meters the area.

06. 秦俑三号坑展示厅外景
07. 秦始皇陵出土文物陈列厅南外观

对页
04. 秦俑一号坑展示厅外景
05. 秦俑二号坑展示厅外景

06. Pit NO. 3 outside view
07. Qin Mausoleum relics exhibition hall south side view

Opposite page
04. Pit NO. 1 outside view
05. Pit NO. 2 exhibition hall mapping

Yan'an Revolutionary Commemorative Museum

延安革命纪念馆

延安，是屹立在黄土地上的历史丰碑，它曾经是中华民族始祖黄帝及其部族的故乡；延安，是中华民族重要发祥地和中国革命圣地，毛泽东等老一辈无产阶级革命家在这里生活和战斗了 13 个春秋，奠定了中华人民共和国的基石；延安，是孕育光照千秋延安精神的发祥地，在中国的革命和建设中发挥了巨大的精神动力作用。

延安革命纪念馆坐落在延河之滨的王家坪。这里依山傍水，山上苍松翠柏，脚下延水滚滚。王家坪大桥宛若一道彩虹横跨延河两岸，将纪念馆与城区连在一起。迈步穿过大桥，是宽阔平坦的纪念馆广场，广场上绿草茵茵，花团锦簇，中央矗立着一尊通高 16 m 的毛泽东铜像。铜像双手叉腰，目视东方，浩气凌霄。主楼上方正中央高悬着郭沫若题写的“延安革命纪念馆”七个金色大字，在阳光下熠熠生辉。

延安革命纪念馆占地面积 158 675 m^2，广场面积 38 000 m^2，主体建筑面积 29 853m^2，高为 19.5 m，局部高达 28 m。内部“延安革命史”陈列面积 7 030 m^2，序厅 1 008 m^2，半景画馆 630 m^2，临时展厅 2 009 m^2。该馆主体共三层，其中一层主要用于游客服务、停车、存放文物和后勤保障；二层中间为序厅，东边为临时展厅；二层西侧和整个三层为陈列展厅。

Yan'an, is the historical monument standing on the loess plateau, which was once the Chinese first ancestor Huangdi and tribal homeland; Yan'an, is an important birthplace of Chinese nation and Chinese revolutionary sites, Older generation of proletarian revolutionaries such as Mao Zedong lived and fought in the 13 years, laid the cornerstone of the People's Republic; Yan'an spirit is the birthplace light nurtured future generations. Yan'an spirit of revolution and construction in China has played a huge role in the spiritual power .

Yan'an revolutionary commemorative museum is located in wangjiaping by Yan river., which near the river besides the mountain. Pines and cypresses are upon the hill, the water of Yan river rolls under the feet. Wangjiaping just like in a rainbow bridge across the strait Yan river, which linked the commemorative museum and the city together. Moving across the bridge, is a broad flat Memorial Square, crossed green lawn, and blooming flowers, stands a 16m high statue of Mao Zedong in the center. Bronze hands rested, visual orient, majesticly. Main building on the side is hunging on the center of the Guomoruo's "Yan'an Revolutionary Commemorative Museum" seven golden Chinese characters, shining in the sun.

Yan'an Revolutionary Commemorative Museum covers an area of 158,675 square meters. Square area is 38,000 square meters. Main floor area is 29 853 square meters, high as 19.5 meters, locally up to 28 meters. Internal "Yan'an Revolutionary History" display area 7,030 square meters, 1,008 square meters. Half view art house is 630 square meters, 2,009 square meters of temporary exhibition hall . The main museum has three floors, The first floor is mainly used for tourist services, parking, storage heritage and logistics. The second floor is preface hall, and temporary gallery on the east ; west of the second floor and the hole third floor are exhibition hall.

01

01. 序厅

对页
02. 博物馆全景鸟瞰
03. 馆后半圆弧
04. 侧立面

01. Prelude hall

Opposite page
02. Bird-eye's view of the museum
03. Semi-circle in the back of the museum
04. Side Facade

02

03

04

Han Yangling Museum

汉阳陵博物馆

汉阳陵为西汉景帝刘启的陵园，位于咸阳原东端，地跨今陕西省咸阳市渭城区和西安市泾阳、高陵县三县、区，属全国重点文物保护单位。博物馆于2004年8月开始动工建设，2006年3月31日建成并对外开放。

帝陵外藏坑位于帝陵陵城之内，围绕封土四周呈放射状分布。为了不影响陵园的历史环境风貌，建筑设计从分析文物及地层的关系入手进行构思。鉴于现在的地面距汉代坑口地面距离为6~7 m，具有可利用的空间高度，设计者采用了全地下建筑的设计方式，建筑顶部覆土植草。主入口避开司马道正面，置于东阙门北阙台西侧，尽量淡化建筑与环境的冲突，化建筑于无形，保持陵园原有的历史环境风貌和自然景观。

根据外藏坑的空间分布，建筑师考虑了独特的建筑结构选型、结构布置与施工方案。设计采用大跨度预应力钢筋混凝土梁门式结构，跨内吊装，将需要覆盖保护的10条外藏坑划分为两个区域，较短的15~21号坑为一区，梁沿东西布置，跨度31 m；

Situated at the east-most of Xianyang Highland and spanning Weicheng District in Xian'Yang City, Jingyang and Gaoling County in Xi'an City, Han Yangling Mausoleum Complex is the burial place of Emperor Jing in the Western Han Dynasty and is the place designated under state protection. It began to build in August, 2004 and was opened to the public on 31st, Mar, 2006.

Lying inside the emperor's graveyard, the burial pits were arranged around the emperor's mound in a radial pattern. To maintain the ancient historical environment of the mausoleum, the design of the protection hall has taken the relations between the heritage and the stratum into consideration. Since the depth between today's floor and the pit is about 6 to 7 meters, which can be used during the design of the construction, the whole building was built underground with green plants on the top. The main entrance is located to the west of northern platform of the East Gate-Tower, keeping away from the Sacred Way. The clash of the modern building and the ancient environment is lightened greatly with the building invisible and the original historical and natural environment of the mausoleum maintained.

According to the space location of the burial pits, the designer adopted the unique construction material, structure arrangement and construction scheme. The pre-stressed concrete beams were adopted, which divide the 10 burial pits need to be protected into two sections. The beams in section

01

01. 展陈细部

对页
02. 总平面
03. 博物馆主入口
04. 入口坡道

01. Detail of Exhibition

Opposite page
02. Master plan
03. Main entrance of the museum
04. Entrance ramp

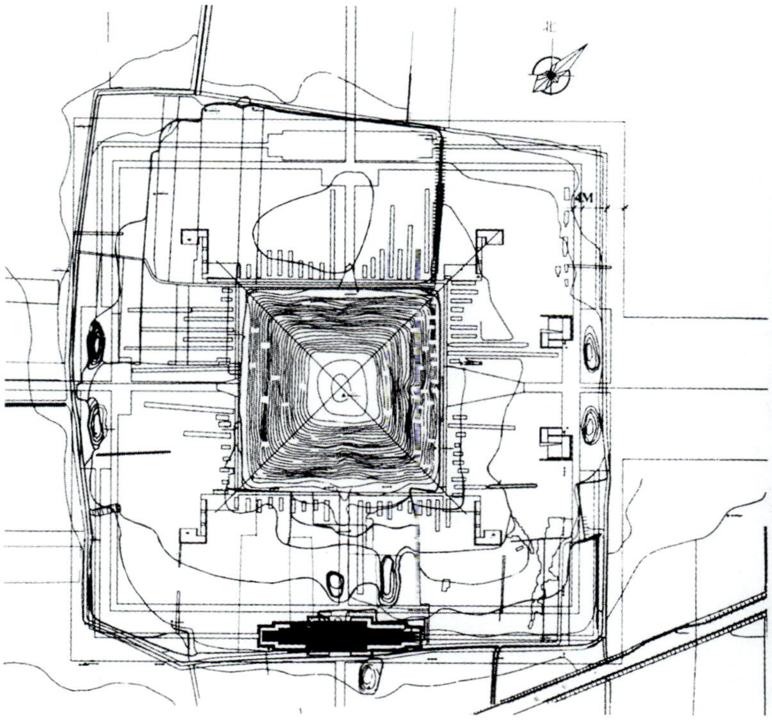

02

03

较长的 12~14 号坑为二区，梁沿南北布置，跨度 15 m 。

根据外藏坑文物遗址坑的分布状况及类型，建筑师设计了丰富的参观流线。从入口开始，建筑师在长达 500 m 的展线中，通过引、停、绕、跨、靠、观等手段，配合灯光照明，引导游客从不同方位和视角观察文物，造成跌宕起伏、一回三折的游览历程。

one including No 15 to 21 pits run from west to east, spanning 31 meters, while those in section two including No 12 to 14 pits run from north to south, spanning 15 meters.

Based on the location and classification of the heritage, the designer created a rich flow. Along the 500-meter exhibition line from the entrance till the end, the architect guides the visitors to observe the heritage from different angles through such means as introducing, stopping, by-passing, crossing, reclining, observing with the decoration of lights, creating an undulating and varied mind's journey.

04

05

05. 玻璃合围的外藏坑遗址大厅

对页
06. 玻璃合围的外藏坑遗址大厅
07. 帝陵外藏坑保护展示厅

05. Relics hall of outer pit surrounded with Glass

Opposite page
06. Relics hall of outer pit surrounded with glass
07. Exhibition hall of outer pit of the emperor's mausoleum

06

07

Xi'an Museum

西安博物院

西安博物院位于西安南郊朱雀大街中段全国重点文物保护单位小雁塔保护范围内，距明代西安城墙 1 km，距西安咸阳机场逾 40 km。西安博物院占地面积 245 亩(约 160 000 m^2)，由三大部分组成：唐代建筑小雁塔及其附属的明清荐福寺古建筑群，现代化博物馆和荐福池园林景区。

小雁塔建于唐景龙年间（707—710 年），本名"荐福寺塔"，是唐代长安著名佛寺荐福寺的佛塔。小雁塔历经 1 300 年后，保持了唐代初修时的原貌，是中国早期密檐式佛塔的代表作。现荐福寺除小雁塔为唐代建筑外，其余殿宇均为明清及近代所建，基本保持了明英宗正统（1436—1449 年）年间重建后形成的格局，规模严整，朴素大气，与小雁塔共同构成了一处完整的文化遗产。

现代化的博物馆位于小雁塔近旁，采用与小雁塔呼应的仿唐建筑风格，建筑面积 16 050 m^2，展厅面积逾 5 000 m^2，公共面积 3 000 m^2，地下一层，地上二层，建筑高度 27 米，在 2004 年被评为西安未来十大标志性建筑之一。馆内收藏有文物藏品 12 万件（组），以青铜器、玉器、金银器、陶瓷器和石刻造像为特色。这些藏品大部分出自西安地区的重要遗址。博物馆设有

Xi'an Museum located in the ancient architecture protection area of southern city, Zhu que avenue. It is 1 kilometer from Ming City wall, 40 kilometers from Xian Yang International Airport. Xi'an museum covers the total area of 245 mu (160,000 square meters), composed by three parts: Small Wild Goose Pagoda in Tang dynasty and their subsidiaries: The Small Wild Goose Pagoda and Ming & Qing architecture complex protection area, the modern museum and the garden.

The Small Wild Goose Pagoda was built during the years of Jinglong (707-709A.D.), original called "Jian fu Temple Pagoda". It was the pagoda for the Buddhist temple--Jian fu, which is very famous in Tang Chang'an city. The Pagoda is a brick structure with multiple eaves displaying a high level of artistry; it is the classical Tang style. Other architecture complex was rebuilt in Ming dynasty.

The modern museum located near the Small wild goose pagoda, covering area is 16,050 square meters with exhibition area of 5,000 square meters and public area of 3,000 square meters. The building is 27 meters high with one floor underground and two floors overground. It has been regarded as one of the ten symbolic buildings in Xi'an city in 2004. There have over 120,000 pieces or sets of collections in the storage room of Xi'an museum, mainly include Browns, Jade, Gold & Silver, Pottery & Porcelain and Stone figures. Most of the collections are unearthed from the main sites in Xi'an area. The museum includes the basic exhibition— "Ancient capital Xi'an", special exhibition-- Buddhist figures, Jade wares, seals and ancient calligraphies. Moreover, the modern scientific technology is also used to add some entertainment by people's involvement.

01

01. 大厅莲花吊顶

对页
02. 博物馆鸟瞰
03. 隋唐长安城模型
04. 博物馆基本陈列"古都西安"展区

01. Lotus-like Ceiling

Opposite page
02. Bird-eye's view of the museum
03. Model of chang'an city in sui and tang dynasty
04. Permanent exhibition area "ancient capital Xi'an"

02

基本陈列“古都西安”，专题陈列佛教造像、玉器、印章和明清书画，并且辅以多媒体互动展示等多种科技互动展示。

荐福池园林景区以占地 7 000 m² 的荐福池为主体，点缀唐三彩叠置伎青铜塑像及喷泉广场、“长安古乐陈列馆”、拴马桩等民间石刻，辅以亭、台、水榭等古典建筑，形成特征鲜明的园林景观区。

Scenic area of Jian fu Temple covers the area of 7,000 square meters. The main part is the Jian fu Pool, others include Tri-colored pottery acrobats brown statue, fountain square, “the music in ancient Chang’an exhibition hall”, hitching post, pavilion, waterside pavilion etc, which forms distinctive features of the landscape area.

03

04

Museum of Gansu Province

甘肃省博物馆

甘肃省博物馆展览大楼位于兰州市七里河区西津西路的南侧，平面呈 T 字形，总建筑面积 20 331m²，抗震设防为九度，楼面荷载为 4kn/m²。主展楼为五层，两翼三层，层高 5.6m ～ 7.2m。全楼共有大小展厅 18 个、学术报告厅 1 个。展厅总面积 13 732m²，展线总长 2 610.4m。中路 7 个展厅为全封闭式、用人工采光、带中央空调系统；两翼 12 个展厅采用自然和人工采光相结合。整栋展楼配置有较先进的火灾自动监控报警灭火系统、安全防盗监控报警系统、空调系统、智能控制变配电系统和讲解、电化教育、广播、闭路电视、网络及计算机管理系统。

该楼土建工程由甘肃省建筑设计研究院设计、装修装饰工程由北京弘高建筑装饰设计工程有限公司和深圳三鑫特种玻璃技术工程有限公司设计，土建工程由甘肃省第一、二建筑工程公司施工，装饰工程由甘肃同泰环艺公司、上海健尔斯公司、广东装饰总公司、兰州兰穗公司施工。2006 年 11 月底竣工，大楼装修荣获全国建筑工程装饰奖、甘肃省建设工程飞天金奖。

The exhibition building of the Museum of Gansu Province, arranged in T-shape, is located on the southern side of West Xijin Road in Qilihe District of Lanzhou City, with the gross floor area of 20,331 square meters and seismic resisting to degree 9, combining administrative office, warehouse, exhibition in to one intelligent building and providing a variety of modern facilities and equipped with overall video service, large screen and touch-screen. The internet service presents the relics, studies, exhibitions of the museum to the whole world.

The exhibition building has five floors, with three-floor wings on both sides and 5.6 ～ 7.2 meters height. One lecture hall and 18 exhibiting halls of different sizes occupying 13,732 square meters exhibiting area and 2,610.4-meter-long exhibiting line totally. Seven exhibiting halls in the mid-way part of the exhibition building are completely occluded and equipped with artificial lighting and central air-conditioning while twelve exhibiting halls on two sides with both natural lighting and artificial lighting and the whole building with relatively advanced fire monitoring, alarming and precaution system; security and monitoring and warning system; air-conditioning system; intelligent control system of power conversion and power distribution; electronic guiding equipment; electronic educating system; broadcasting apparatus; closed circuit TV; internet and computer administrating facilities.

01

01. 博物馆旧影（摄于 1959 年）

对页
02. 外景
03. 彩陶展厅一角
04. 丝绸之路展厅

01. Historical scene of the museum (in 1959)

Opposite page
02. Exterior
03. A corner of the exhibition hall of colored pottery
04. Exhibition hall of the silk road

02

03

04

Qinghai Museum

青海省博物馆

青海省博物馆成立于 1986 年 9 月，原馆址是利用建于 20 世纪 40 年代的马步芳公馆，在西宁市城东区为民巷 36 号。该馆占地近千亩，房舍 300 余间，建筑面积逾 6 000 m^2，包括住宅、办公、花园及仓房等。大门门楼上用汉白玉镌刻“馨庐”二字，系国民政府主席林森所题。部分建筑的内外墙面用汉白玉砌成，以示豪华富丽。该建筑群为三进院落，结构严谨，布局合理，具有中西式相融的风格，是典型的近代回族庭院式建筑。后因其建筑、展厅等不能适应快速发展的西宁市文化发展和博物馆事业要求，20 世纪末，在日本友人小岛遼次郎先生支持下，青海省人民政府在西宁市城西区另选新址，建了新馆。

2001 年，青海省博物馆整体迁入位于西宁市城西区新宁广场东侧的新馆中。该馆占地 17 000 m^2，主楼建筑面积 20 800 m^2，共 5 层，内设主、侧展厅 9 个，面积 9 146 m^2，另有库房、文保、会议厅等设施，是集展区、办公区、文物库房为一体的具有现代化功能的大型博物馆。其整体建筑采用传统的中轴对称手法，正面为 16 根方柱，显得大气、庄重。主屋顶为重檐歇山顶，四角配有四座亭子遥相呼应，传统建筑与现代手法运用自如，具有浓厚的民族、地方特色，散发着强烈的时代气息，与馆前宽广雄伟的新宁广场交相辉映，构成了西宁古城一道亮丽的风景。

Established in September, 1986, Qinghai Museum was located at the former site of Ma Bufang's Residence (former warlord) built in 1940s with its address of No. 36 Weimin Lane, Eastern District, Xining. The museum covered an area of about 1,000 mu with more than 6,000 square meters of floor area, including residence, administrative offices, gardens and barns, etc.. On the gate tower was Xinlu (Cottage of Virtue) inscribed, the autograph of Lin Sen, the chairman of the former Government of the Republic of China. Parts of the external walls were paved with white marbles, a symbol of richness and luxuries. With the precise structure and well-balanced arrangement, the architectural cluster which is representative of modern Muslim courtyard-style was designed with three compound entrances incorporating both Chinese and Occidental styles. However, the buildings and exhibition halls gradually became incapable of fulfilling the needs of Xining's culture and museum industries which were in full swing. At the end of the 20th century, with the financial support of Mr. Ryojiro Kojima, a friend from Japan, new Qinghai Museum was built in the Western District of Xining and the site was carefully selected by the People's Government of Qinghai.

In 2001, Qinghai Museum was relocated to the present new site at the eastern side of Xinning Square, Western District of Xining. The new one is a large-sized museum which embodies modern functions with the integrated exhibition area, administrative area and warehouses of cultural relics. The new museum covers an area of 17,000 square meters and the five-storeyed main building has a floor area of 20,800 square meters with 9 main/side exhibition halls 9,146 square meters. Furthermore, there are also warehouses, preservation facilities of cultural relics, conference halls and other facilities in the new museum. The whole architecture is built by traditional axial symmetry methods: There are 16 square columns in the front with splendid and magnificent views and the main roof is a double-eaved one with four pavilions on four corners echoing with each other. The perfect combination of traditional and modern architectural techniques has made the museum rich in local and ethnical characteristics yet remained quite modern. Adding radiances and beauty to each other, Qinghai Museum, along with the splendid Xinning Square, has become a picturesque scenic spot of Xining, an ancient city with a profound history.

对页
01. 新馆外景
02. 老馆外景
03. 新馆展厅一角

Opposite page
01. Exterior of the new museum
02. Exterior of the old museum
03. A corner of the exhibition hall of the new museum

01

02

03

Museum of Xinjiang Uygur Autonomous Region

新疆维吾尔自治区博物馆

新疆维吾尔自治区地处祖国西北边陲，不仅幅员辽阔、历史悠久，更有特殊的自然地理环境，地上地下文物古迹极为丰富。新疆维吾尔自治区博物馆位于首府乌鲁木齐市西北路 581 号，是自治区唯一的省级综合性博物馆，是全疆最大的文物和标本收藏保护、科学研究和宣传教育机构。1953 年成立博物馆筹备处，1958 年始建，1959 年正式成立。

原馆初步设计为农业展览馆，为“山”字形平房建筑，主体中心高达 30 m，拱形圆顶具有浓郁的民族特色。建筑面积 8 427 m^2，陈列展厅 7 000 m^2。中央大厅和建筑门厅两侧全部饰有图案精美的石膏花饰，主体建筑门厅两侧墙壁上有巨幅“飞天”瓷砖壁画。“新疆维吾尔自治区博物馆”馆名是朱德委员长 1959 年来新疆视察工作时题写的。

新馆于 2005 年 9 月竣工开馆，建筑面积 17 228 m^2，其中展厅面积 7 136 m^2，地下一层，地上二层，主体高 18.5 m，玻璃穹顶高 29.5 m。建筑平面呈“一”字形对称布局，具有浓郁的西域风格和新疆地方特

Xinjiang Uygur Autonomous Region is located in the northwest edge of the motherland. It is vast with a long history, and it has especially geographical environment. It is rich in the cultural relics above and under the ground. Museum of Xinjiang Uygur Autonomous Region lies at No.581 of Northwest Road of Urumqi, the capital city. It is the only comprehensive museum which has the provincial level, and is the largest collection of relics and specimens for protection, scientific research and public education institution. The Preparatory Office was established in 1953, and the museum was built in 1958 firstly, and was formally established in 1959.

The original preliminary design was the Agricultural Exhibition Hall, which was in the shape of a hill and was a ranch house. The center of main building reached up to 30 meters, with a strong ethnic characteristic of the arched dome. The floor area is 8,427 square meters with an exhibition hall of 7,000 square meters. The central hall and the two sides of the construction hall were both decorated with fine plaster floral pattern, and there were two huge murals named "Flying" on both sides of the ceramic-tile wall. "Museum of Xinjiang Uygur Autonomous Region" was inscribed by Chairman Zhu De who made an inspection in Xinjiang in 1959.

The new museum was completed in September of 2005. Its floor area is 17,228 square meters, with an exhibition area of 7,136 square meters. There is a basement underground and two stories on the ground.

01

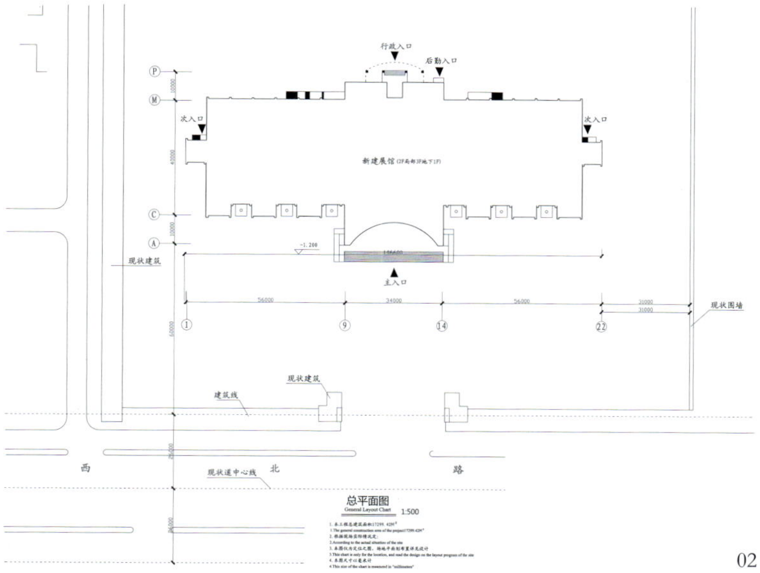

02

03

02. 总平面
03. 冬日下的老馆

02. Master plan
03. Old museum in winter

对页
01. 新馆外景

Opposite page
01. Exterior of the new museum

色。外墙饰面为天然大理石和白色大理石镶嵌，室内主要为磨光花岗岩地面和墙壁。主体建筑内有八根方形立柱，一层中央有新疆地貌圆形沙盘，在二层有环形走廊，可以俯瞰一层中央圆形沙盘。一、二层两侧墙壁雕刻有新疆历史、民俗文化风格画廊雕像。

Height of the main building is 18.5 meters, and that of the glass dome is 29.5 meters. The architecture surface is like the Chinese character "一". The whole architecture is rich in the style of the Western Regions and Xinjiang local features. External walls are the natural marble and white marble inlaid in it , and in the interior room, the main floor and walls are polished granite, of which the main building there are eight square columns. There is a layer of the central circular sandbox which displays the landscape of Xinjiang, in the second floor there are circular corridors, from central layer of the second floor visitors can overlook the circular sand table, on the first and the second sides of walls are engraved with folk culture gallery and statue.

04

05

06

07. 剖面
08. 正立面
09.《新疆民族风情陈列》展厅

对页
04. 新馆二层回廊
05. 新馆中央大厅
06. 一层中央大厅俯视

07. Section
08. Facade
09. Exhibition hall (natural condition and social Customs of the nationalities in Xinjiang)

Opposite page
04. Ambulatory on 2nd floor of the new museum
05. Central hall of the new museum
06. Top-view of the central hall on ground floor

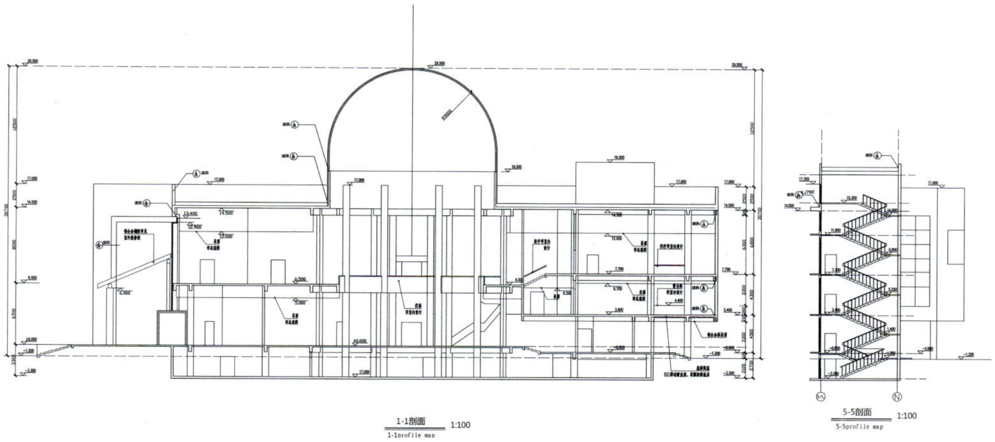

07

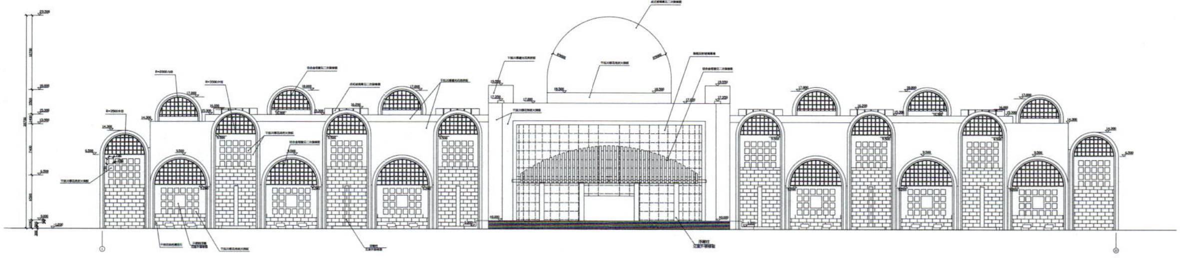

08

09

Hong Kong Museum of History

香港历史博物馆

香港历史博物馆建成于 1975 年，1998 年迁至现址。历史馆新馆毗邻香港科学馆，同样是由巴马丹拿建筑及工程有限公司按照美国的威拿 · 庄信先生的建筑概念设计而成。博物馆建筑群前方有宽敞广场，中央平台相连，在周边高楼密集的环境中，辟出一片舒展的文化空间。两馆外型、色调和物料互相协调，空间结构亦一体融和，形成一个博物馆组合。科学馆以直线为主，历史馆则采用弧线设计，馆内展厅均宽敞无柱。

历史博物馆建筑面积 17 000 m²。其中 7 000 m² 的常设展陈“香港故事”为博物馆的核心，展览的陈列设计由加拿大的雅克安德烈顾问公司负责。 博物馆的建筑设计演绎出香港历史远古至今的层层推展。由参观者步入展厅南端的起点开始，圆拱形的上盖向北一层比一层加阔增高，覆盖的展区则一段较一段宽阔深邃。整个建筑结构空间根据展览陈列而规划，参观至半途转往上层之际，设有可透视馆内庭园的休憩区和餐厅，方便休闲小歇。

Hong Kong Museum of History was established in 1975 and moved to its new building in 1998. It shares its site with the neighbouring Hong Kong Science Museum. Both two museums were built by Palmer & Turner Architects and Engineers Ltd. based on the architectural concept of Mr E. Verner Johnson of USA. The raised plaza, courtyard and public park spaces offer a welcome relief for the dense urban surroundings. The museum complex was developed to form a unified whole of a harmonious appearance, similar materials and colour scheme. While the Science Museum is more rectilinear, the Museum of History has more curves, which allow for long-span exhibit spaces free of interior columns. The museum comprises a floor area of 17,500 square metres. The heart of the premises is 7,000 square metres “Hong Kong Story” permanent exhibition which layout is designed by Andre and Knowlton Associates Ltd of Canada. The architectural design for the museum expresses the unfolding nature of Hong Kong's dramatic growth over time. The vaulted forms of the building begin with smallest in the south, corresponding to the beginning of the visitor's path through the exhibit areas. They get increasingly larger in height and footprint as they progress to the north. The interior volumes of these forms are appropriately sized to fit the soaring exhibits, which also get progressively more spacious and grand as they move forward. The visitor flow through the exhibit sequence is set up architecturally so that there is a comfortable break at the halfway point, where visitors

01

02. “香港故事” 常设展上层展览厅平面图
03. 香港历史博物馆外貌

对页
01. 香港历史博物馆前的广场

02. Floor plan of upper galleries of “The Hong Kong story” permanent exhibition
03. Exterior of the Hong Kong museum of history

Opposite page
01. Piazza of the Hong Kong museum of history

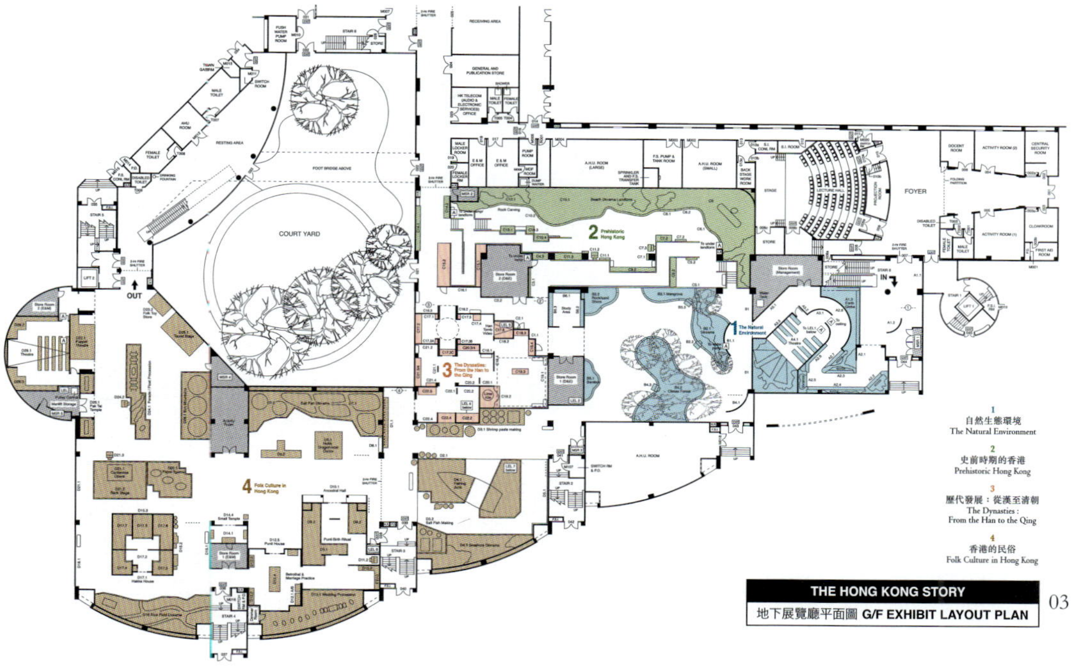

03

02

03

香港历史博物馆的建筑特色以参观者的全面体验为大前提。从构思到完成，建筑的设计与展览的布局，一直同步并进，环环相扣，处处融合，互为彰显。

transition to the top floor of exhibits, with a Cafe and Resting Area overlooking an internal courtyard.

The architectural design focuses on the total visitor experience. There is a seamless overlap of the exhibit design and architectural design, which each reinforcing the other. During the entire design process, through the final construction, there was constant coordination between the exhibit needs and building needs.

04. "香港故事"常设展内，以场景展示长洲太平清醮，包括包山、戏棚、飘色、舞狮和道坛等。
05. "香港故事"常设展内，以场景展现数千年前先民的生活。

对页
06. "香港故事"常设展下层展览厅平面图

04. The diorama in "The Hong Kong story" galleries demonstrating the lives and activities of early inhabitants.
05. The reconstruction of the festive activities of the Taiping Qingjiao ceremony in cheung chau, including the bun mountains, the cantonese opera theatre, the parade, the lion dance, and the taoist altar, in "The Hong Kong story" galleries.

Opposite page
06. Floor plan of lower galleries of "The Hong Kong story" permanent exhibition

04

05

THE HONG KONG STORY
閣樓展覽廳平面圖 M/F EXHIBIT LAYOUT PLAN

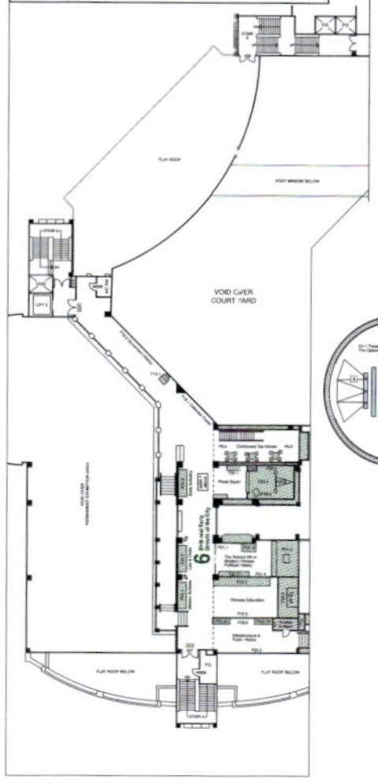

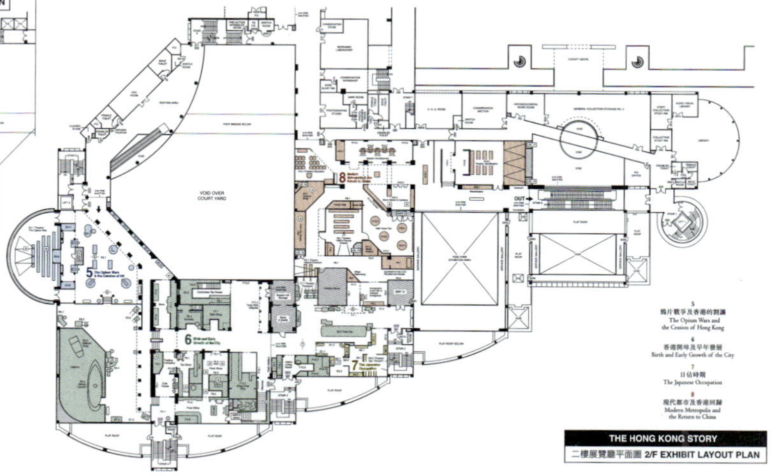

THE HONG KONG STORY
二樓展覽廳平面圖 2/F EXHIBIT LAYOUT PLAN

Maritime Museum of the Macao Special Administrative Region

澳门特别行政区海事博物馆

澳门特别行政区海事博物馆于 1987 年开幕，现展览大楼在 1990 年开始正式启用。海事博物馆位于澳门半岛西南端，世界文化遗产妈阁庙的左前方，据说其所在地就是几百年前葡萄牙人首次到达澳门时登岸之处。澳门的历史亦由此开始改变，由一个宁静的小渔村，发展到今天的国际知名的都市。

海事博物馆占地约 2 400 m²，展览大楼面积约 1 000 m²，建筑构想来自航海大发现事迹，由乘船渡海开始。因此，海事博物馆的外形独特，远看就好像一艘帆船，锚泊在码头旁；内部的设计，也跟船舱相像，无论布局、楼梯、窗户、管道等，都让人有置身船内的感觉。

博物馆楼高三层，以中国、葡萄牙及澳门地区的航海活动为主线来划分专题展区，地下展厅主要介绍澳门渔民社群的生活情况，以渔民家庭及澳门内港鱼栏特色建筑为展厅的背景；一楼介绍中国及葡萄牙的航海历史，配以古船及中葡航海图为设计主线；二楼介绍航海技术知识及海上交通

The Maritime Museum of the Macao Special Administrative Region was inaugurated in 1987 and the existing exhibition building was officially opened in 1990. Situated on the southwest tip of the Macao peninsula, the Maritime Museum lies on the front left side of the A-Ma Temple, a World Cultural Heritage site, where the Portuguese first landed in Macao several hundred years ago and the transformation of the history of Macao also started from here. Macao has developed from a small peaceful fishing village into today's renowned international city.

The Maritime Museum covers an area of 2,400 square metres, with an exhibition area of about 1,000 square metres. Since its architectural design was inspired by the Age of Discovery which began with sailing across the sea, the façade of the Museum looks like a vessel anchored by a wharf from a distance, and is very unique. The interior also resembles a ship cabin, making people feel like they are inside a vessel, in terms of layout, staircases, windows and pipes.

The Museum is a 3-storey building, consisting of different thematic exhibition areas, which depict Chinese, Portuguese and Macao maritime activities respectively. Constructed with a background featuring fishermen's families and the distinguished architecture of the fish market in Macao's interior harbour, the ground floor mainly exhibits the life of the fishermen communities in Macao. The first floor is dedicated to introduce Chinese and Portuguese maritime history, integrating ancient ships as well as Chinese and Portuguese nautical charts in the design. The exhibition on the second floor focuses on the technology and maritime transport. The decoration of this area was based on a ship cabin and the starry sky, with the wavy carpets and the mast in front of

01

01. 露天茶座之遮阳篷

对页
02. 一层平面
03. 立面
04. 博物馆全景

01.The shade structure of the esplanade

Opposite page
02. Plan of 1st Floor
03. Facade
04. Panorama of the museum

工具，以船只的驾驶舱及星空为布置的蓝图，配合驾驶舱前眺望阁楼的波浪地毯和船帆，墙壁上刻有中式罗盘及西式罗经花，使人置身于扬帆起航的情景之中。

除展览区域外，博物馆还设有水族馆及欧陆色彩的露天茶座供游客享用。

the cabin overlooking the Mezzanine Floor, and walls carved with Chinese compasses and western compass roses, allowing people to indulge themselves in the picture of setting sail on a voyage.

Besides the exhibition areas, the Museum also houses a gallery with aquariums and an esplanade of European style, which is open to visitors.

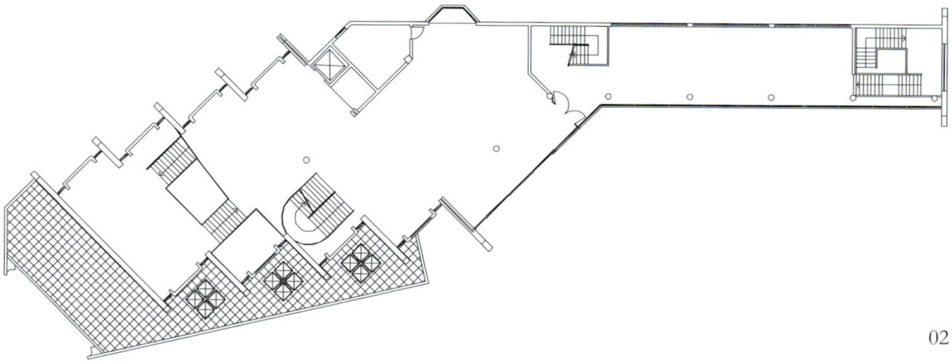

02

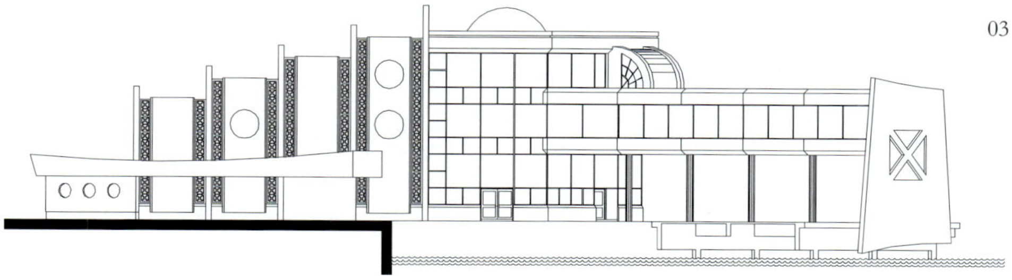

03

04

06. 博物馆内貌
07. 位于博物馆二楼的海事技术展览厅

对页
05. 位于博物馆一楼的海事历史展览厅

06.The interior aspect of the maritime museum
07.The exhibition gallery on the second floor focuses on the marine technology and maritime transport

Opposite page
05. The exhibition gallery on first floor focuses about the chinese and portuguese maritime history

Macao Museum of Art

澳门艺术博物馆

澳门艺术博物馆于 1999 年 3 月 19 日开幕，坐落于澳门地标之一的澳门文化中心。澳门艺术博物馆建筑面积 10 192 m²，设计充满现代都市风格，很容易拉近与参观者的距离。

澳门文化中心由艺术博物馆大楼和剧院大楼两部分组成。这两座建筑物由一个艺术广场和景观公园连接起来。艺术广场是整体设计的主线，在首层的中间形成一个宽敞的户外空间。

澳门艺术博物馆楼高五层，正面布局非常严谨，由中间主门廊分隔两旁并列的银灰色建筑组成。穿过博物馆的正面门廊，到达剧院大楼，顶部是一巨型银灰色金属飞檐。飞檐微微向海面倾斜，仅覆盖部分建筑。这道鲜明的飞檐，形态轻快而色调沉稳，仿佛中国书法之笔触，在附近众多矩形建筑物中散发耀眼光芒，别具一格。

澳门艺术博物馆各楼层平面均采用开放式设计。馆的背面为明净的落地玻璃，可尽览艺术广场的景色。建筑师刻意利用玻璃幕墙将各楼层通道和中央广场连接起来，

The Macao Museum of Art was inaugurated in 19th March 1999 and located in Macau Cultural Centre which is one of Macau landmarks. The Museum Building covers a total area of 10,192 square meters, providing the urban framework and allowing proximity between the visitors and the exhibition rooms.

The architectural complex of the Macao Cultural Centre includes two main buildings, the Macao Museum of Art and the auditoria, which are connected by a central plaza and surrounded by gardens. The plaza creates the main lines of force that organize and structure the general composition and give rise to a large intermediate, outdoor space at the first floor level.

Macao Museum of Art is a 5-storey building, and its facade, very controlled, composed of juxtaposed planes with horizontal openings, is interrupted only by the main portico. Through an entrance portico, the building containing the auditoria, closed of with an enormous and expressive steel roof consisting of a curve sheet of varying thickness sloping towards the sea, partially covering the building. The lightness and the solidity of this curved bright surface, like a calligraphic gesture in suspension, stand out against the background and distinguished the whole complex from the massive orthogonal volumes of NAPE.

Macao Museum of Art is open design for all floor plan. There are bright French windows on the back with overlooking the are plaza. The architect deliberately connected the access on each floor with the central plaza by use of glass curtain wall and along the same direction of the central plaza to each exhibition hall which have the visual effect in connection between flowing spatial, access, exhibition area and outdoor space as well as more harmonious.

对页
01. 博物馆夜景
02. 平面

P344-345 页
03. 南宗北斗——董其昌诞生四百五十周年书画特展
04. 金相玉质——清代宫廷包装艺术展
05. 巨匠心影——毕加索版画展

Opposite page
01. Night scene of the Macao Museum of Art
02.Floor Plan

P344-345
03.The light of the southern school: special exhibition on the 450th anniversary of the birth of dong Qi Chang
04.Lustrous syntheses: The art of imperial packaging of qing dynasty
05.Reflections of the master's mind: an exhibition of picasso's prints

01

沿着与中央广场同一方向的通道到达各楼层展厅，使博物馆的流动空间、通道、展区和广场户外空间在视觉形成强烈的连接效果，令内外空间更为和谐。

澳门艺术博物馆中间主门廊面临马路，是整座建筑物的正门。除了正门的石阶入口，参观者亦可经由面向海旁的坡道和行人天桥进入澳门艺术博物馆。澳门艺术博物馆各层功能如下：藏品库、艺术工作坊、多媒体图书馆和小型展览厅。

Macao Museum of Art, to the west, giving onto the street, forms the main entrance to the whole complex. Besides the main entrance stairs, the museum is accessible from the garden by means of monumental stairways, from the side facing the sea by an elevated walkway and by a ramp. The museum building house include the storage area, a large workshop area, a library and a small-scale exhibition gallery.

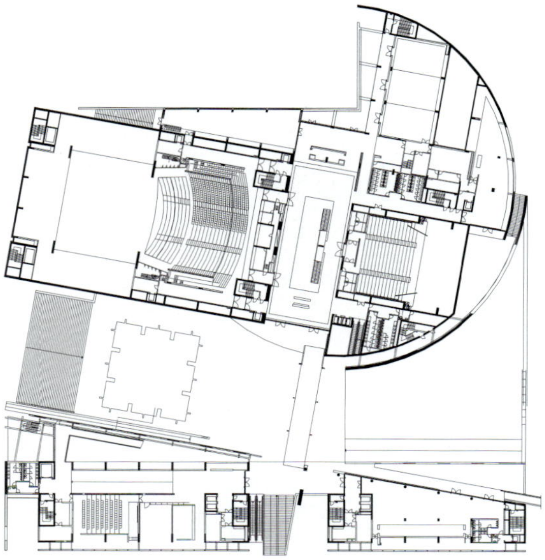

02

03

A EMBALAGEM DE
ARTIGOS DE
Articles for
LÚDICO
文玩
A EMBALAGEM DE
The Packaging of
MATERIAL DE
Stationery and
ESCRITA E DE
Treasures
OBJECTOS VALIOSOS

04

05

宗教經典與法物包裝
A ARTE DE
The Packaging of
EMBALAGEM DE
Religious
ESCRITURAS RELIGIOSAS E
Scriptures and Implements
UTENSÍLIOS
EMBALAGEM DA
Painting and
PINTURA E
Calligraphy
CALIGRAFIA
EMBALAGEM DA
The Packaging of the
PINTURA
Palace Painting of the
PALACIANA DA
Qing Dynasty

图书在版编目（CIP）数据

中国博物馆建筑：英汉对照／BIAD传媒《建筑创作》杂志社主编.—天津：天津大学出版社，2010.11
ISBN 978-7-5618-3758-0

Ⅰ. ①中… Ⅱ. ①建… Ⅲ. ①博物馆…建筑设计－简介－中国－汉、英 Ⅳ. ①TU242.5

中国版本图书馆CIP数据核字（2010）第098691号

策划编辑 金 磊 韩振平
责任编辑 韩振平 郭 颖
版式设计 胡珊瑚

出版发行 天津大学出版社
出 版 人 杨 欢
地 址 天津市卫津路92号天津大学内（邮编：300072）
电 话 发行部：022-27403647 邮购部：022-27402742
网 址 www.tjup.com
印 刷 北京雅昌彩色印刷有限公司
经 销 全国各地新华书店
开 本 250mm×285mm
印 张 28.6
字 数 394千
版 次 2010年11月第1版
印 次 2010年11月第1次
定 价 298.00元